SpringerBriefs in Applied Sciences and Technology

Ethical and Legal Issues in Biomedicine and Technology

Series editor

Giovanni Boniolo, University of Ferrara, Ferrara, Italy

More information about this series at http://www.springer.com/series/13628

Giovanni Boniolo · Virginia Sanchini
Editors

Ethical Counselling and Medical Decision-Making in the Era of Personalised Medicine

A Practice-Oriented Guide

 Springer

Editors
Giovanni Boniolo
Dipartimento di Scienze Biomediche e
  Chirurgico Specialistiche
University of Ferrara
Ferrara
Italy

and

Institute for Advanced Study
Technische Universität München
Munich
Germany

Virginia Sanchini
Department of Experimental Oncology
European Institute of Oncology (IEO)
Milan
Italy

ISSN 2191-530X   ISSN 2191-5318   (electronic)
SpringerBriefs in Applied Sciences and Technology
ISBN 978-3-319-27688-5   ISBN 978-3-319-27690-8   (eBook)
DOI 10.1007/978-3-319-27690-8

Library of Congress Control Number: 2015960428

Printed on acid-free paper

This Springer imprint is published by SpringerNature
The registered company is Springer International Publishing AG Switzerland

Contents

About the Book

This book offers an overview of the main questions arising when biomedical decision-making intersects ethical decision-making. It reports on two ethical decision-making methodologies, one conceived for the patients, the other for the physicians. It shows how patients' autonomous choices can be empowered by increasing awareness of ethical deliberation, and at the same time, it supports healthcare professionals in developing an ethical sensitivity, which they can apply in their daily practice.

The book highlights the importance and relevance of practicing bioethics in the age of personalised medicine. It presents concrete case studies dealing with cancer and genetic diseases, where difficult decisions need to be made by all the parties involved: patients, physicians, and families. Decisions concern not only diagnostic procedures and treatments, but also moral values, religious beliefs, and ways of seeing life and death, thus adding further layers of complexity to biomedical decision-making. This book, which is strongly rooted in the philosophical tradition, features non-directive counselling and patient-centeredness. It provides a concise yet comprehensive and practice-oriented guide to decision-making in modern healthcare.

The Plan

Giovanni Boniolo

Abstract We introduce the issue of the Ethical Counselling, showing its usefulness to improve patients' awareness on the ethical choices they may be asked to do
whenever clinical options raise ethical dilemmatic situations.

Patient first: i.e., patients' quality of life and the imbricated quality of their decisional process.

Case 1

*Giovanna (38) is a single mother of two children (5 and 3). At the age of 25, she
was diagnosed with breast cancer but thanks to the early diagnosis and good
response to treatment she recovered. There is a known history of cancer in
Giovanna's family: her father died of colon cancer, her paternal aunt had breast
cancer, and two of Giovanna's cousins also developed tumours. As doctors suspected, Giovanna has been tested positive for TP53 mutation. This particular
germline mutation is associated with Li-Fraumeni syndrome, which indicates the
increased risk (up to 85 %) of developing tumours in early adulthood: bone and
soft-tissue sarcoma, premenopausal breast carcinoma, leukaemia, brain cancer,
and adrenocortical carcinoma. Some of these cancers also affect children.
Giovanna knows that the father of both children does not have any history of
cancer in his family and TP53 mutation, being a dominant autosomal disorder,
implies that there is 50 % chance that each of her children inherited TP53 mutation
from her. Giovanna is thinking about testing her children for TP53 mutation, but
her religious mother keeps telling Giovanna that she should rely on God's will.
Meanwhile, Giovanna wants to protect her children from having to face cancer but
is not sure if undergoing the testing is the right thing to do. She is also feeling*

G. Boniolo (✉)
Dipartimento di Scienze Biomediche e Chirurgico Specialistiche, University of Ferrara,
Ferrara, Italy
e-mail: giovanni.boniolo@unife.it

G. Boniolo
Institute for Advanced Study, Technische Universität München, Munich, Germany

uneasy about managing the information in case one or both of her children test positive. What will she be expected to do with such information? Will she be obliged to communicate the results to the children? If so, when and how should it happen?

Case 2

Claudio (29) is a cancer survivor, who had his semen collected and frozen before undergoing chemotherapy when he was 15 years old. His parents chose a private cryostorage facility and paid all the costs. Now, they cannot wait to have a grandchild. Claudio and his partner Francesca (32) have been trying to conceive a baby without success. Recently, they were seen by a fertility specialist who explained that intracytoplasmatic sperm injection (ICSI) would be the best option with an estimate for a successful pregnancy ranging up to 37 %, noting that research has been limited and these numbers come from very small, single site studies. Moreover, choosing this option would expose Francesca to numerous medical procedures (retrieving the eggs, transferring the embryos to her uterus, etc.). Claudio is not sure if it is fair on Francesca to expose her to all these interventions when success rates are just about one-third. In addition, Claudio is also concerned that his cancer might recur and he might die before his child is independent. Should Claudio do everything possible to become a parent?

Case 3

Anna (27) comes from a small village in the south of Italy. She is married to Matteo, and they have a 3-year-old son. Their attempt to have a second child resulted in two miscarriages. However, Anna just discovered that she is again two months pregnant. Soon after, she felt something strange under her arm and went to see her doctor. The diagnosis was dreadful: oestrogen-positive breast cancer! Anna was offered three therapeutic paths which could be followed in order to treat her disease: (1) standard treatment which is not compatible with pregnancy; (2) surgery followed by adapted chemotherapy which is compatible with foetal development but has lower response rate to treatment; (3) just surgery, postponing all other treatments until after the delivery of the baby. Matteo is categorically against any choices that could compromise Anna's survival. What should Anna do?

Case 4

Julie (41) is married for ten years to Philip. They are both lawyers and only recently decided to have kids. Because of her age, and in order to increase her chances of pregnancy, Julie decides to undergo assisted reproduction, more specifically in vitro fertilisation (IVF). IVF is successful, two embryos are implanted in utero and both attach, giving rise to a twin bicorial pregnancy. After fifteen weeks of gestation, Julie undergoes amniocentesis at the local department of medical genetics. Amniocentesis is a prenatal test that is routinely offered to women who are at an increased risk for bearing a child with birth defects or chromosomal aberrations, like Down's syndrome. The outcome is unusual: although the twins are males, one of them has the normal karyotype (XY), while the other has an extra male (Y) chromosome. This condition, known as YY syndrome, does not give rise to *severe clinical implications. However, the medical geneticist explains to Julie that, with a certain likelihood (around 50 %), affected individuals may develop mild language and learning impairments, in particular they may show the symptoms of attention deficit hyperactivity disorder (ADHD). With respect to IQ levels, however, XYY children do not show any significant difference with children having the XY karyotype. The physician also reassures Julie that the studies that in the past associated the YY syndrome with severe psychiatric disorders leading to criminal attitudes are now dismissed by the medical community as scientifically flawed. Julie is a positive and confident person and, in agreement with her husband, decides she will not submit the kids to postnatal testing in order to identify who of them is affected by the YY syndrome. Her motivations are as follows: "I do not want to discriminate among them. If they should want to know, they will test as adults". Is Julie and Philip's choice not to test the twins ethically acceptable?*

Case 5

Veronica, 45, is married with two children. Her friend Sara just got a positive result from her mammogram and the following biopsy confirmed a breast cancer for which she underwent surgery and chemotherapy facing severe side effects. Thus, Veronica decides to ask her doctor whether she shall go for a mammography. The physician performs a clinical breast examination, a physical examination to check for lumps or other changes that suggest a possible cancer lesion, and finds nothing strange. Veronica has no family history of breast cancer, but still she was 37 years old at the birth of her first child and she is now 45. The physicians points out the potential benefits and harms of breast cancer screening test in Veronica's clinical case by presenting her with statistical data on mortality reduction, the probability of false-positive results and overdiagnosis, combined with the patient's anamnesis and the absence of sign or symptoms. Veronica and her husband are facing a dilemma. On the one hand, they are worried about possible life-threatening conditions and they want to be sure everything is fine even though she is asymptomatic. Therefore, Veronica might to undergo a mammography screening claiming, "I know about possible harmful outcomes, but I rather be safe than sorry. I'm sure my children would agree". On the other hand, Veronica is very touchy and sensitive about medical procedures and she is worried about under-taking useless and potentially harmful medical practices. Therefore, she might decide not to undergo a mammography screening claiming, "I was impressed by Sara's story, but why should I check if I'm feeling well?".

These are just some exemplary cases in which patients in clinical settings must navigate not only between *clinical* but also between *ethical* options. Indeed, the complexity of all these and similar cases stems from the fact that each of them raises two kinds of interrelated yet distinct questions: questions concerning the clinical side of the cases, and questions concerning patients' values, moral sensitivity, and ethical perspectives.

Of course, not every clinical decision raises relevant ethical questions. However, as both doctors and patients are well aware of, routine clinical practice is fraught with difficult situations in which what *should be done* at the clinical level appears unclear because it is unclear what *ought to be done* at level of the moral options at stake. Whenever this happens, one can say that one is facing not only a clinical but also an *ethical decisional problem*, which sometimes could become an *ethical dilemma*. A dilemma is a decisional conflict occurring within a single agent whenever he/she must decide between two or more mutually exclusive courses of action, so that selecting one option necessarily results in discharging the other. The peculiar feature characterising *ethical* dilemmas is that the reasons that the agent provides in favour of one of the two alternatives are specifically *moral* reasons, that is, reasons concerning moral principles and values.[1]

But how do people decide when they face an ethical decisional problem? To answer this question, it is worth looking at a series of recent empirical findings about human moral decision-making.[2] For our purpose, the results of these empirical studies can be summarised in four general statements, which we co-opt hereafter among the grounding assumptions of our view about how ethical decisions in clinical settings should be approached. First, *humans are morally flexible*. This means that rather than deciding and judging on the basis of a well-constructed moral theory, most people tend, instead, to decide and judge according to a vaguely structured moral framework and the requirements of the social context they are part of. This almost unstructured moral framework that human beings seem to possess constitutes a sort of normative toolkit of decisional heuristics the subject makes use of every time a moral decision has to be taken. Second, *this "moral toolkit" is almost always unconsciously assumed, and it usually reflects the influence of many biographical variables* (personal history, culture, tradition, etc.).[3] Third, *emotional reactions depend on complex moral judgements*. Experimental moral philosophers have also pointed out that the first automatic, apparently purely emotional answer to a moral demand is actually rooted in normative principles ("I must …") or in rightful claims ("I have the right to …") belonging to that moral framework. Finally, *one's morality may not be negotiable*. As other researches have emphasised, people often possess moral values and principles that they consider as unquestionable or sacred.

Once considered together, these assumptions support the view that, (i) in face of the same ethical dilemma, different persons may take different decisions because they rely on a diverse, often unique set of moral tools, heuristics, and guiding principles; (ii) this moral toolkit is often unconscious, and therefore, despite

---

[1] For a deeper introduction on the concept of ethical dilemmas see McConnell T (2014) Moral Dilemmas, The Stanford Encyclopedia of Philosophy, http://plato.stanford.edu/archives/fall2014/entries/moral-dilemmas.

[2] See, for example, Bartels DM et al. (forthcoming) Moral Judgment and Decision Making. In: Keren G, Wu G (eds) The Wiley Blackwell Handbook of Judgment and Decision Making, Wiley, Chichester (UK); Knobe J (2010) Person as scientist, person as moralist. Behav Brain Sci. 33:315–29.

[3] It is important to note that the complexity of the moral world and the flexibility of human nature was already known and discussed since the Greek philosophers and Aeschylus, Sophocles and Euripides' tragedies (by the way, also in contemporary age these features have been stressed in many writings coming form totally different traditions such as F.M. Dostoyevsky's *Demons* and P. Levi's *The Drowned and the Saved*).

possessing it, human beings are not aware of it; (iii) despite the initial certainty of our intuitive response, we may revise our decisions whenever they conflict with some parts of our moral framework; and (iv) there are ideas that people are unlikely to question and change.

These assumptions constitute the core elements of what we define as *Personal Philosophy*. This expression refers to that wide set of more or less deep, coherent, and justified metaphysical, methodological, religious, political, esthetical, ethical, etc., beliefs, assumptions, principles, and values that an agent possesses and that characterises in a unique way how he/she approaches the world and his/her life. In other words, Personal Philosophy could be considered as the "conceptual and value-laden window" from which any individual starts reflecting in order to make judgments, to make choices, and to act.[4]

The importance of taking into account the very personal and ethical dimension of the decision-maker—what we have just called Personal Philosophy—appears particularly relevant in the era of the so-called *personalised medicine.*[5] Personalised medicine can be broadly defined as "the tailoring of medical treatment to the individual (especially genetic and epigenetic) characteristics, needs and preferences of the patient during all stages of care, including prevention, diagnosis, treatment and follow-up".[6] Even if the term is sometimes used interchangeably with "precision medicine", "stratified medicine", and "targeted medicine", the focus is always on providing the right treatment at the right dose to the right patient at the right time. However, despite its importance and novelty, personalised medicine has been much more appreciated by researchers and clinicians, rather than by patients. A possible reason lying behind this fact might be the fact that patients appear much more interested in the dimension of the "care" broadly considered—that is, in the act of taking care of a human being conceived as a whole—rather than in the practice of "disease treatment" in a strictly technical sense.[7] Nevertheless, in the broader debate on personalised medicine, this *personal patient care* aspect has not received much attention, an aspect to which, instead, we want to give a proper consideration. Personal Philosophy within the ethical decision-making process could be considered as the equivalent of personalised medicine within the clinical decision-making process. To be more precise, the possibility of letting the patient

---

[4]We privilege this term instead of the classical *Weltanschauung* (*World view*) because the latter belongs to a specific German philosophical tradition; see Dilthey W (1907). Das Wesen der Philosophie. Marix Verlag, Wiesbaden 2008; Jasper K (1919) Psychologie der Weltanschauungen, Verlag von Julius Springer, Berlin. Another very close concept would be *Weltbild* (*World conception*), which, nevertheless, has had less biography then *Weltanschauung*.

[5]See: European Science Foundation. Report on Personalised Medicine: http://www.esf.org/coordinatingresearch/forward-looks/biomedical-sciences-med/current-forward-looks-in-biomedical-sciences/personalised-medicine-for-the-european-citizen.html.

[6]U.S. Food and Drug Administration (FDA) (2013) Paving the way for personalized medicine: FDA's role in a new era of medical product development. October 2013. Report.

[7]See: Cornetta K, Brown CB (2013) Perspective: Balancing personalized Medicine and Personalized care. Acad Med 88(3): 309–313; Van Heist A (2011) Professional loving care: an ethical view of the healthcare sector. Ethics of care series. Volume 2. Peeters, Leuven.

decide in line with his/her Personal Philosophy seems nothing but the ethical equivalent of providing the patient with a tailored treatment. Both personalised medicine and Personal Philosophy appear, thus, as significant dimensions contributing to the development and improvement of the personal patient care ideal.

Summing up, many times *clinical decision-making* goes hand in hand with *ethical decision-making*. However, sometimes, the difficulty of choosing between incompatible ethical options may give rise to an ethical dilemma, which can then affect also the decisions to be taken at the clinical level. Every time this occurs, patients tend to rely on their Personal Philosophy in order to decide which option ought to be chosen in any given situation. Since the decision on which clinical option ought to be pursued may depend on a choice that is made at the level of the moral decisional problem at issue, doctors willing to put into practice a personalised approach to medicine and care cannot avoid facing patient's ethical dilemmas through the exploration of their own Personal Philosophy. Moreover, Personal Philosophy could be considered, as the equivalent of personalised medicine in the ethical dimension. As the latter requires that the patient be provided with an ever more specific and tailored treatment, the former assumes that clinical choices involving ethical decisions are taken in relation to the patient's Personal Philosophy.

This set of considerations suggests the need to have proper tools that help in comprehending and coping with ethical dilemmas in clinical settings. The necessity of having such tools can be also justified in the light of preventing some potential risks, both on the side of patients and on the side of clinicians. In particular, two kinds of risks might be identified here.

- *The risk for the patient* to fall into what could be thought of as an *ethical decisional paralysis*. Indeed, there are people able to decide by themselves, but there are also people who could find themselves in the situation as described by Seneca to Lucilius, the procurator of Sicily: "No man by himself has sufficient strength to rise above it; he needs a helping hand, and someone to extricate him".[8]
- *The risk for the clinician* to consider himself/herself as ready to deal with patients' ethically dilemmatic situations only on the basis of their own Personal Philosophy. As all human beings, clinicians too possess their specific and individual Personal Philosophy. However, as we will better explain in the third chapter of this book, the mere fact of being clinicians does not legitimate them to impose it on patients whenever extremely serious patients' life choices are on the stage. On the contrary, clinicians should be trained to face them in order to properly recognise them and thus to avoid imposing, even unwillingly, their personal standpoints in a directive *manner on patients*.

In this book, we use the label "Ethical Counselling" to refer to a comprehensive methodology through which it is possible to mitigate the problems that might arise

---

[8]Seneca LA (62-65) Moral Letters to Lucilius, Letter 52, 2.

when both patients and doctors face ethical decisional problems in clinical settings, as well as meet the patient's need of receiving a more comprehensive and personalised care. More specifically, in our view,

> Ethical Counselling is a dialogic activity implementable in the cases in which clinical decisions involve ethical issues. It always involves the presence of two actors–the ethical counsellor and the counselee–, and it has two different purposes. On the one hand, by clarifying and investigating patients' Personal Philosophy, it assists them (with or without their relatives) to break through their ethical decisional paralysis in clinical settings and to choose the option more in line with their ethical sensitivity. On the other hand, it trains clinicians to properly examine the ethical dilemma that their patients are facing, in order to go beyond their commonsensical and intuitive moral understanding and to avoid the dangerous conviction that their own Personal Philosophy is better than patients' one.

In particular:

- Ethical Counselling is not a form of psychological support, even if, being directed towards patients beliefs, it crosses with some psychological aspects.
- Ethical Counselling is not aimed at providing solutions to ethical dilemmas. Rather, it is conceived as a "way of cleaning the windows" in order to have a deeper comprehension of them. This, in turn, is done in order to allow patients to be more aware of their choices and clinicians of the situation they are addressing.
- Ethical Counselling is not a mandatory tool that should be imposed on patients. Instead, it might be proposed to them and they might use it, in case they feel the necessity for it. Differently, some nudges to undergo an ethical counselling process seem to be necessary for clinicians since, as said, an intuitive moral understanding does not prove to be enough in dilemmatic cases involving patients.
- Ethical Counselling is a tool by means of which the counsellor serves the patient or the clinician (according to two different methodologies, as we will explain in the methodological part of this book) to reflect and thus to tame the first intuitive answer. Note that this *reflection slot* between the rising of the dilemma and reaching its solution allows also for a critical examination of the representation of the clinical event the patient is living. Usually, human beings have a starting representation of the event at issue, and they proceed to make a moral evaluation based on it. Thus, if an ethical counsellor introduces time to discuss such a representation, this could be better considered, or reconsidered, and, probably, a better moral evaluation and solution could come out.

If the Ethical Counselling is a personalised tool focused on patient's Personal Philosophy, the ethical counsellor is someone who, by using his/her knowledge and expertise and through a properly constructed dialogue, serves the patient to ethically reflect on *his/her* way of living and thinking, in order to let *him/her* take important moral decisions concerning diagnostic possibilities and therapeutic treatments that might have a great impact on his/her life and on the life of his/her relatives.

The purpose of the Ethical Counselling is to allow for a reflection slot in which the first automatic and emotional answer to a dilemma is posed under review, as it happens also to the starting representation of the clinical event. Moreover, it allows to individuate the patient or clinician's value ranking in order to understand whether there are undisputable or non-negotiable moral values and principles. However, as it will be explained in the next chapters, even if the clinician, being a human being, has his/her Personal Philosophy, this should not come first onto patient's one.

In the following chapters, the ideas just mentioned will be properly explained and justified, in order to offer a theoretically informed guide on Ethical Counselling.

In Chapter "Ethics Consultation Services: The Scenario", we try to give an idea of the state-of-the-art concerning ethical consultation services and the so-called *philosophical counselling*. By doing this, we could also prepare the ground to make the typicality of our approach to Ethical Counselling more perspicuous. In particular, as Chapters "Ethical Counselling for Patients" and "Ethical Counselling for Clinicians" highlight, our approach is based on two methodologies (one designed for patients and one designed for clinicians) that are historically rooted in our philosophical tradition. The methodology we propose in the case of Ethical Counselling for patients is strongly grounded in the Aristotelian practical philosophy, which could be considered as the first decision-making theorisation. Instead, the methodology we propose in the case of Ethical Counselling for clinicians follows the presentation of the *status quaestionis* with which the Medieval scholars began their *disputationes*, that is, the technique designed to clarify and solve a problem: exactly the tool necessary for clinicians to clarify the ethical problems met by their patients.

By pondering on these two methodologies, the reader could easily grasp that they are intended also to offer the *reflection slot* mentioned above. In fact, it allows for a better analysis of the initial Personal Philosophies, of patient's representations of the clinical event, and of the reasoning by means of which evaluations and solutions are justified. Note that, from this point of view, our way of proposing Ethical Counselling is deeply entrenched with critical thinking in the field of ethically dilemmatic decisional situations.

This, so-to-say, more theoretical part is followed by three chapters in which we present three topics that have to be extremely familiar to any ethical counsellor: communication, emotions, and probability.

Handling communication properly (Chapter "Nocebo and the Patient–Doctor Communication") is one amongst the main skills required for structuring an appropriate patient-physician relationship, but it is also vital whenever a service of Ethical Counselling is implemented. In the chapter devoted to it, we do not dwell upon how the communication should be made: a certainly relevant question that we would invite to be enquired separately. Differently, we discuss an aspect that should be known by the ethical counsellor, that is, its neurocognitive counterpart and its possible effects on the patient's brain. Informing patients about potential side effects of treatments and interventions can induce nocebo effects which refer to adverse events related to negative expectations and anticipations. Neurobiological and preclinical studies have shown that nocebo effects result from negative expectations, previous experiences, and clinical interactions. In particular, a proportion of adverse events are due to patients being informed about potential adverse events of medication. Disclosures and the manner in which information is delivered can contribute to producing adverse effects.

Emotions (Chapter "Reasons and Emotions") should be also taken into account in the Ethical Counselling process. Not only do they permeate the representation of the clinical event as constructed by the patient and not only should emotional narratives find room during the counselling, but they also have to be tamed. As we will discuss in the chapter concerning the counselling for patient, a relevant point known since a couple of thousands of years ago concerns the fact the a good decision-maker has to be characterised by what the ancients called *sophrosyne*, that is, temperance: the virtue of ruling emotions in order to make the decision as rational as possible. This does not mean, of course, that we want to discard emotions from the ethical counselling process, but that by means of the *reflexion slot* that it allows, the first emotional answer might be tamed in order to arrive at a (at least partially) detached solution. Surely, the degree of detachment depends on the individual patient, but the ethical counsellor should work in the direction of considering patient's emotions by trying to govern them, also in order to prevent them from becoming harmful. In this chapter, we discuss also why we consider reason-giving as an essential component for moral deliberation and thus of Ethical Counselling. In particular, we shall consider: (i) the role of emotions with respect to reason-giving; (ii) why we should rely on reason-giving in ethical choices; (iii) how reason-giving makes explicit patients' Personal Philosophies; (iv) why reason-giving is always an open-ended process.

Probability (Chapter "The Centrality of Probability") is at the core of many ethical decisions encountered in clinical choices in the age of personalised medicine. *Carrier testing* for cystic fibrosis tells us whether the individual has one copy of a mutated CF transmembrane regulator (CFTR) gene. In this case, the individual tested positive will have 50 % of having a child who is a carrier of the same condition and has to decide his/her reproductive strategy. If also the partner is a healthy carrier, the potential child will have 50 % chances of being a healthy carrier and 25 % of having the condition. How to decide about becoming parents? *Predictive and presymptomatic testing* could provide the information about whether an asymptomatic individual will surely or only probabilistically develop a specific pathological condition. If an asymptomatic individual is tested positive for the mutation of Huntingtin gene, they will surely suffer from Huntington's disease: a pathology, which has no treatment so far. Thus, what should this individual do? How should he/she manage his/her life plan? To be positive to genetic variants for type 2 diabetes tells you that you have from 10 to 35 % chances to develop the pathology. What does this mean? Is it worth doing the test? Probability has also to do with the *survival rates* that tell us the probability over time, for example, to die of a particular disease since it has been diagnosed or treated. But what does this mean? Let us recall that for an individual (be it an asymptomatic one or with an already overt pathology), it is not relevant at all to know a statistical probability but to know his/her particular fate. Thus, he/she has to understand the probabilistic information given by the clinician but, first of all, the clinician has to know the meaning of probability. And, of course, an ethical counsellor cannot be unprovided with such knowledge.

After a first part devoted to these methodological features, the second part deals with some of the most important questions an ethical counsellor could face (and thus should know) in the context of personalised medicine.

We begin (Chapter "Genetic Testing and Reproductive Choices") by offering an introduction for the management of the ethically controversial issues arising from the relationships between reproductive choices and genetic testing. Rather than going into the details of each and any argument present in the bioethical literature on the topic, we prefer to provide a conceptual road map that should serve the ethical counsellor for correctly framing the case he/she may be asked to confront with.

Then, (Chapter "The 'Right-not-to-Know'") we present a very related issue, that is, that one concerning the ethical controversies resulting from the exercise of the so-called right-not-to-know. By problematising the substantive conflicts at the basis of patient's decision to waive some health-related information, the chapter provides a normative map for orienteering the practice of ethical counselling in the face of these claims. In particular, we discuss a number of ethical and legal aspects of the right-not-to-know in the context of Julie's case introduced above.

Also the following chapter (Chapter "Incidental Findings") is in somehow connected with the previous two, since it deals with the so-called incidental findings. Various medical tests are routinely performed in medical practice to establish or confirm a diagnosis and prescribe the right treatment. In some cases, the results of a medical test can reveal a previously undiagnosed condition, which is not related to the current medical condition and the original purpose of the test. Such results are referred to as "incidental findings" and have sparked a significant debate on whether all of these findings or some, and which ones, should be reported back to the patients. We discuss the implications and the management of the incidental findings in the clinic and the debate that surrounds them, to offer a method for reaching an ethically desirable outcome on the case-by-case basis.

We then proceed with the questions regarding the ethical choices occurring in the field of oncofertility (Chapter "Oncofertility"). In this chapter, we present the most pressing ethical concerns surrounding fertility preservation for cancer patients by giving an overview of the arguments supporting and rejecting fertility sparing interventions.

Finally, we deal with the issue regarding overdiagnosis (Chapter "Overdiagnosis"), one of the main concerns in screening context, which we particularise on cancer. By overdiagnosis, we refer to the detection of a tumour, which does not constitute a substantial hazard to the patient, but it is treated as it was. In cancer clinical practice, indeed, overdiagnosis represents a possible harmful outcome of screening to which both patients and physicians should pay careful attention. We will explain how ethical counselling could provide aid in understanding probabilities and emphasising patients' values in order to reach an informed decision about whether to undergo screening tests.

We conclude with some remarks on what the ethical choice could be in the light of what has been said in the previous chapters.

Acknowledgments We would like to thank B. Bonanni, L. Chiavari, G. Curigliano, I. Feroce, and F. Lalatta, the members of the now closed Biomedical Humanities Unit at the Department of Experimental Oncology (Istituto Europeo di Oncologia, Milano).

Part I
Methodology

Ethics Consultation Services: The Scenario

Alma Linkeviciute and Virginia Sanchini

Abstract This chapter aims to provide a review of the state of the art of the ethical consultation services. In particular, we present a map of the most important traditions, discussing the different conceptions and the different roles they give to the counsellors. Then, we show the place of our ethical counselling methodology within this scenario.

Keywords Ethics consultation · Moral case deliberation · Ethical counselling · Philosophical counselling

The need for what could be generally defined as *ethics consultation*[1] emerged together with the increasing development of technological progress within the field of biomedicine. In particular, technological advances of medical equipment made it possible to keep terminally ill patients alive by using machines for maintaining their vital functions. As much as this technological progress provided hope, it also created new ethical problems. Amongst them, one of the most famous cases occurred in Seattle in the early 1960s. The ethical dilemma concerned the allocation of scarce availability of life-saving renal dialysis machines. There were just few machines available and hundreds of people facing death in a matter of weeks. Some physicians felt that they should not be the only ones involved in these difficult decisions and that more competences would have been necessary to fully evaluate the complexity of the situation. The need for interdisciplinarity as well as for sharing decisional burdens led to the establishment of the first ethics committees,

---

[1]We use 'ethics consultation' as a general term covering all different types of ethics support services, which can be offered to patients and clinicians.

A. Linkeviciute
Dipartimento di Scienze della Salute, University of Milano, Milan, Italy
e-mail: alma.linkeviciute@ieo.eu

A. Linkeviciute · V. Sanchini (✉)
Department of Experimental Oncology, European Institute of Oncology (IEO), Milan, Italy
e-mail: virginia.sanchini@ieo.eu

G. Boniolo and V. Sanchini (eds.), *Ethical Counselling and Medical Decision-Making in the Era of Personalised Medicine*, SpringerBriefs on Ethical and Legal Issues in Biomedicine and Technology, DOI 10.1007/978-3-319-27690-8_1

which were composed of clinicians and non-clinicians, in order to represent all the societal interests involved in the decisions to be made (Tapper 2013).

Since then, numerous attempts have been made to address ethical issues surrounding medical decisions and, roughly speaking, at least two major traditions could be individuated. The first, known as *clinical ethics consultation* (henceforth CEC), has been mainly practiced in the USA. It is characterized by the effort to solve conflicts through a dialogue between all the stakeholders involved (Fiester 2015). The second tradition, more prominent in Europe, might be collected under the umbrella name of *moral case deliberation* (henceforth MCD). Here, the emphasis has been put on the conceptualizing process preceding the solution rather than on the problem-solving dimension, hence providing healthcare professionals with a space for discussion and idea-sharing (Dauwerse et al. 2014). Moreover, increasingly, ethics consultation is being delivered on different levels ranging from decisional support to an individual patient to helping various stakeholders in resolving ethical disagreements, and also advising institutions and/or governing bodies on ethical decision-making. Our focus will mainly be on individual patients and clinicians requiring ethics consultation rather than institutions.

1 A Map

After the first committee was created in the context of renal dialysis, several committees started to appear all over the USA. However, since at the very beginning these committees did not have any legal standing, the optimal scenario would have been to leave the contingent definitive solution to the courts. Unfortunately, difficult ethical decisions were often surrounded also by legal uncertainty. Moreover, since there were neither standard indications nor guidelines, the committee members had to rely only on their conscience and creativity in establishing points of reference to back up their decisions.

Nevertheless, the establishment of the first committees in the USA was inevitably followed by the formation of professional organizations, which, on the one hand, institutionalized the ethics consultation activity and the ethics consultant role and, on the other hand, provided a more concrete space for people to communicate, exchange ideas and learn from each other. In early 1998, the American Society for Bioethics and Humanities (ASBH) emerged after the consolidation of already existing organizations that were working in bioethics and ethics consultation.[2] Soon this organization stated the core competences of ethics consultants, including personal qualities and cultural skills that were thought to be necessary to carry out the consultation process in a proper manner. Moreover, the ASBH recently expanded its purpose and proposed the more comprehensive term of *healthcare ethics*

---

[2]These predecessor organizations were the Society for Health and Human Values (SHHV), the Society for Bioethics Consultation (SBC) and the American Association of Bioethics (AAB).

consultation.[3] It also introduced the first code of ethical responsibilities, which was a robust step towards regulation, quality assurance and control of CEC services in the USA.

Following the ASBH, some institutions started to appear also in Europe. For example, in 2001 the UK Clinical Ethics Network (UKCEN) was established with the scope of coordinating communication amongst ethics committees, organizing educational events and offering ethical advices to healthcare professionals. In 2005, the Cochin Hospital Clinical Ethics Centre was instituted in Paris and then, in Amsterdam, the European Clinical Ethics Network (ECEN) (Molewijk and Widdershoven 2007). The major difference between the ASBH and its European counterparts lies in the approach they endorse and propose. The ASBH rules on how CEC service should be organized and structured, whereas UKCEN and ECEN picture themselves as supporting and guiding healthcare professionals, but they do not aim to establish concrete and binding practice requirements or issuing practice licences to potential ethics consultants. In other words, the ASBH actively seeks to become a regulatory body for CEC services, while European organizations might be considered as primarily focused on education and training of the stakeholders involved.

Differences between the two approaches can be better illustrated by considering the national guidelines framing clinical ethics practices in different countries (Gaucher et al. 2013). In the American setting, ethics consultants are expected to participate in a principled ethical resolution to resolve conflicts and avoid turning into courts, while, for example, British ethics committees, under recommendation of the Royal College of Physicians, are pictured as offering ethics support and advice. In France arose what have been then called *Espaces éthiques.* According to this model, physicians are advised to pursue ethical reflection regarding troubling issues with the help of the ethics committees, but any initiative of sharing decision-making responsibility amongst ethics committees and physicians is discouraged (Gaucher et al. 2013). Depending on the complexity of the case and the implications attached to the solution required, especially in the context of CEC, individual consultant, small group or a committee might serve best (Aulisio et al. 2000). Moreover, also resources' availability and institutional policies appear to have a role in defining whether ethics consultants are part of healthcare team or external service providers (Aulisio et al. 2000).

To sum up, there are various models offering ethical assistance in practical decision-making and theoretical reflection in healthcare settings. North America and to some extent the UK are working on implementing the practice standards and regulating the ethics consultation service; the French approach, instead, shapes ethics support as a space to perform theoretical and philosophical reflections

---

[3]"Health care ethics is an umbrella term that encompasses several ethics 'subspecialties' such as organizational ethics, clinical ethics, professional ethics, business ethics, education ethics, and research ethics. These subspecialties overlap, and ethics consultations often involve more than one subspecialty. For example, clinical ethics, organizational ethics, and professional ethics may be involved in a consultation where a physician-in-training must learn when, and how, to disclose a medical error to a patient" (Tarzan 2013, 4–5).

(Gaucher et al. 2013); finally, Dutch model seems to be a self-regulatory and "learning by doing" enterprise (Stolper et al. 2014).

2 Expectations

2.1 Skills and Knowledge

In order to provide a good service, ethics consultants are expected to possess a number of skills to enable them to carry out ethics consultations in a professional manner. The ASBH, for example, has provided a list of core skills, amongst which there are the capacities to identify and analyse the disagreements; retrieve the relevant ethics literature, policies, guidelines and other documents; collaborate with institutional structures; organize formal and informal meetings; report the activities and establish follow up processes; communicate and train; represent the views of different parties; and recognize and find solutions to communication barriers. Furthermore, also some cultural competencies are supposed, in particular knowledge of moral reasoning, ethical theories and concepts, healthcare systems, institutions and clinical contexts, policies, beliefs and perspectives of patients and staff, relevant codes of ethics, professional conduct, guidelines and health law relevant to ethics consultation (Aulisio et al. 2000; Tarzian 2013).

Also UKCEN has released a document providing the summary of skills and competences required for ethics consultation. However, while ASBH focuses more on individual consultants, UKCEN seems to pay more attention to clinical ethics committees and, therefore, its requirements are specifically devised for the members of such committees (Larcher et al. 2010).

However, both ASBH and UKCEN emphasize the importance of character traits and personal qualities, which are considered necessary for successful ethics consultation: tolerance, patience, compassion, honesty, forthrightness, self-knowledge, courage, prudence, humility and integrity (Aulisio et al. 2000; Larcher et al. 2010). Moreover, the ASBH code of ethics for ethics consultants specifies the core responsibilities which include maintaining the competence and integrity, managing conflicts of interest and respecting privacy, contributing to the field, making public statements responsibly and promoting justice in health care (Tarzian et al. 2015).

The ways to acquire knowledge and skills for doing CEC are still debated. The ASBH is increasingly suggesting examination and licensing, while the UKCEN also acknowledges self-directed learning.

2.2 Service Provider's Role

The ASBH recommends an "ethics facilitation" approach (Tarzian et al. 2015), according to which the healthcare ethics consultant is supposed to have two major tasks. Firstly, the ethics consultant should help in identifying and analysing the nature of the problem. Secondly, the consultant should facilitate the building of a "principled ethical resolution" (Gaucher et al. 2013). This means "ensuring that all involved parties' voices are heard, assisting them to clarify their own values, facilitating understanding of factual information and recognition of shared values, identifying and supporting the ethically appropriate decision-maker(s), and ensuring that identified options comport with relevant bioethics, medical, and scholarly literature, and with laws, pertinent institutional policies and current practice standards" (Tarzian 2013). The ethics facilitation approach endorsed by the ABSH rejects the two extremes of "authoritarian" and "pure facilitation" approaches (Aulisio et al. 2000). Indeed, on the one hand, being persuaded of the equal decisional importance of all the parties involved, it discards the idea that the consultant is the privileged decisional "authority" able to offer the most legitimate solution. On the other hand, even if the ASBH argues in favour of an approach based on dialogue and mutual understanding amongst the different parties involved (as required by a facilitation approach), it does not consider consensus-reaching as the only purpose of the consultation (as a pure facilitation approach would require) emphasizing the importance of communication and education. There are two core elements constituting ethics facilitation as interpreted and used by the ASBH. The first is the process for implementing a meaningful communication amongst all the stakeholders involved. The second is the outcome of the consultation presented as a resolution of ethical problem in concordance with existing legal framework (Tarzian et al. 2015).

In the case of MCD, the requirements for the facilitator are not so explicitly defined. Following the idea that facilitators are those who moderate dialogues, the main professional competences deal with moderation and facilitation skills. These very last considerations lead to the persuasion that the facilitator does not necessarily have to be an ethicist. Since the growing interest in MCD has been reported, for example, in the Dutch health care, training programmes for healthcare professionals are being developed to enable them to run MCD sessions themselves without necessarily depending on (external) ethicists. Teaching programmes for MCD include providing the trainees with basic knowledge about ethics, profound knowledge about conversation methods, attitude and skills for easing dialogue and analytical reasoning. However, each trainee is encouraged to find his or her own style of MCD (Stolper et al. 2014).

Hence, even if both CEC and MCD spur for a "facilitation approach", the way in which such an expression has been interpreted differs quite a lot. CEC consultants facilitate inclusive consensus-building processes, by respecting individual rights to live according to personal perspectives and values, but also recognizing that there are boundaries within which decisions have to be taken (Aulisio et al. 2000). These boundaries are practice standards, guidelines, policies and laws. Facilitators within MCD settings have to lead a dialogue amongst healthcare professionals and, in some cases, amongst the other stakeholders involved, in order to discuss moral issues of a particular case, as well as professional competences, institutional and organizational aspects (Dauwerse et al. 2014). Indeed, MCD sessions can focus on concrete moral questions arising from particular situations, but they can be also expanded as to include more philosophical questions like what it means to be a good professional. In this case, the ethics consultant neither gives any substantial advice and nor is asked to justify a specific decision the group has taken. Overall, MCD is thought as a place to have a constructive dialogue on some moral dimensions specifically linked with the situation analysed or generated but not directly connected with it, and the facilitator's role is to make sure that these scopes are reached properly.

3 Approaches

One of the first attempts aimed at proposing a structured tool for carrying out CEC was made by A.A. Jonsen, M. Siegler and W.J. Winslade. In 1980, they drafted what is known today as the *Four Boxes Method*: a pocket guide for CEC. It focused on medical indications, patient preferences, quality of life and other contingencies (Jonsen et al. 2010). This method incorporated the principles of biomedical ethics[4] that, together with the assessment of the facts in question, are used to reach the solution of the case at stake. The four elements proposed by this method should be interpreted in the following way: (i) medical indications concern the evidence supporting therapeutic and diagnostic interventions; (ii) patient preferences indicate the choices expressed by the patient, based both on personal experience, beliefs, values and on the information provided by the physician; (iii) quality of life describes the features of patients life before and after the treatment, focusing on a degree of satisfaction that they experience and on the values regarding their life and their health; (iv) other contingencies refer to the contextual features such as familial, social, institutional, financial, legal setting within which the medical decision has to be taken. The main aim of this method was to help solving practical and concrete dilemmas, starting with problem identification and concluding with possible strategies to manage the problem. This method stressed the importance of using model cases as guides, which can serve as an educative resource even if they do not provide assurance that the actual case in question could be straightforwardly subsumed to them. Integrated Ethics project developed by US Department of Veterans Affairs also implies similar steps (i.e., clarify the consultation request, assemble relevant information, synthesize information, explain the synthesis, support the consultation process).[5]

Usually, CEC and MCD present a protocol that lists a number of questions to be answered and steps to be followed. The classical way to start the ethics consultation

---

[4]Namely autonomy, non-maleficence, beneficence and justice which constitute mid-level ethical theory referred to as *Principlism* (see Beauchamp and Childress 2013).

[5]The guide specifying further questions and offering further guidance for ethics consultants can be found at http://www.ethics.va.gov/integratedethics/.

is by presenting the case and related problem(s) (Steinkamp and Gordijn 2003). This first moment is followed by the gathering of relevant facts and identification of the stakeholders involved. Then, it proceeds with the comprehension of the nature of the problem, the determination of the values compromised, as well as the arguments in favour and against the potential solutions. Depending on the methodology, the concluding part can arrive at a solution, a recommendation, or at an open scenario in which the final choice should be taken by those directly affected by the problem.

If we look back at the very first committee in Seattle, even though its members did not have access to patients' identities, they considered very closely their socio-economic history, when deciding if one of them was likely to comply with the treatment and benefit from it. This idea has been further elaborated in La Puma-Siegler's method in the 1990s. They advocated the need to talk with a patient on his/her views about life and on his/her social history, which in their perspective was as important as doing physical exams and reviewing laboratory test results (Tapper 2013).

Currently a divergence can be observed between services provided to individual patients and organizations such as healthcare institutions. For instance, patients are offered decisional (Chiavari et al. 2015) and psychological counselling (Lawson et al. 2015) which helps to address personal values, emotional issues and ethical aspects that might influence the decision. Assistance for reaching ethical system-level decisions is also offered in a form of a structured tool (Jiwani 2015) presenting a framework for implementing the ethical decision-making process. It provides a 15-steps guide with worksheets, suggested meeting plans and facilitation tips.

4   Our View

Ethics consultants can help patients, families, healthcare teams and hospital administrations to think, reflect and possibly reach a decision on ethical issues that arise in a context of problematic individual or general clinical cases. Allowing time for ethics consultation can help, on the one hand, understanding where the ethical problem lies and what its main causes are, while, on the other hand, assessing the feasible options and the most acceptable solutions.

Depending on national and local structures, general interests and cultural contexts, the approaches and methods might vary. Probably, unique approaches or methods will not ever be applied universally. Nevertheless, the universal feature shared amongst different approaches and circumstances is the recognition that ethics plays an essential role in contemporary health care since it fosters the delivery of a better personalized care and it improves patient's empowerment and wellbeing.

Considering the two mentioned traditions, our view of *Ethical Counselling* echoes more with the MCD approach as it highlights the importance of clarifying and conceptualizing the problem rather than offering a solution. However, in line with CEC, our perspective rejects both an authoritarian and a pure facilitation approach. Indeed, on the one hand, it considers the parties involved as granted with specific proper autonomies that should be respected and, on the other hand, it discards the idea that consensus-building completes ethical reflection. Moreover, it shares some common ground with another tradition, the *Philosophical Counselling*. Philosophical Counselling can be defined as a specific kind of dialogical activity through which, by resorting to philosophical concepts, theories and techniques, a consultant (or a counsellor) helps a counselee (who is considered here, and therefore defined, as a "client") to reflect upon his/her life troubles.[6] It is to note that Philosophical Counselling has been gaining a growing acknowledgement in the clinical ethics field as an alternative to dispute resolution thanks to its potentiality to offer better "closure" of the case where not only resolution is considered relevant, but also the space to look more carefully at the specific concepts and patterns of reasoning shaping the counselee's perspective (Matchett 2015).

Our view of Ethical Counselling (in the two methodologies we are going to present in the next two chapters) focuses on those newly emerging ethical issues specifically characterizing the field of personalised medicine in the meaning here promoted of personalised care, where the comprehension and promotion of patient's Personal Philosophy plays a central role and, even, a more important role with respect to the technical elements of the care. Personalised medicine offers more specific therapeutic interventions and diagnostic tools but not all of them, especially diagnostic, screening and prevention interventions have precisely defined benefit. Diagnostic and preventive interventions can also carry risks—not only medical, but also psychological, social and existential risks. Our aim is to provide a methodology that is both specifically focused on ethical concerns in the light of personalised medicine and better grounded in our philosophical traditions.

Moreover, in conjunction with the different types of issues personalised medicine raises with respect to other more traditional settings of clinical domain, it also differs with respect to timing. To be more precise, the concerns personalised medicine raises do not always require immediate decisions and there is time for reflection and deliberation by both patients and physicians. For instance, testing one's children for gene mutations, which could result in adult cancer, can wait a year or two or even until children are old enough to take such a decision by themselves; fertility preservation for cancer patients might allow the time of few weeks to consider if such option is the right one for a specific patient; reproductive choices by partners, who are known to be mutation carriers, do not have to be made in one day either. Therefore, providing them with reflection slots rather than only with decision-making tools could be a better approach.

Finally, it has been suggested that different methodologies might serve better for different situations involving different issues and concerns (Steinkamp and

---

[6]It was introduced by a German philosopher, G. Achenbach, who, in 1984, published a book containing some lectures he gave on what he defined as a new way of doing philosophy, according to which philosophy has to be interpreted as a practical tool to cope with life and its difficulties (Achenbach 1984).

Gordijin 2003). Some decisions arising from applications of personalised medicine might not be so strongly bound to law like, for example, end of life decisions, compliance with institutional policies or national legislation. In addition, quality of life can also be seen in a broader sense than just ability to perform physical tasks or appreciate intellectual life. It can be affected by stress caused by extensive testing; uncertainty about the interpretation of the results; pressure to exhaust all diagnostic interventions in order to ensure that disease can be prevented in the future. However, we will see that not all diagnostic interventions can ensure prevention (testing negative for TP53 or BRCA1 is not a guarantee that a person will not develop cancer), just like not all preventive interventions will result in clear benefit for a patient (not all cancer patients will be willing or be well enough to make use of cryopreserved gametes or tissues to resume their fertility, just like not all will lose their fertility during treatment).

Therefore, there is a clear need to provide the updated tools for reflection which would enrich the insights to the ethical concerns arising in a light of personalized medicine. And our view tries to meet this need. Although it resonates closely with other methodologies, it contributes to the existing debate in the following manner: (i) it provides a methodological structure for enhanced reflection which combine in a new syntheses core elements of already existing and leading traditions, such as CEC, MCD and Philosophical Counselling; (ii) it offers separate methodologies for patients and clinicians, as a consequence of the recognition that they play two very different roles in the decision-making process; (iii) it has specific focus on new ethical issues arising in personalised medicine considered as personalised care where new and creative approaches to ethics support might be required for serving best the needs of patients and clinicians; and (iv) finally, as it will be shown in the next two chapters, it is well grounded in our cultural heritage.

References

Achenbach G (1984) Philosophische Praxis. Vorträge und Aufsätze. Dinter, Colonia

Aulisio MP, Arnold RM, Youngner SJ (2000) Health care ethics consultation: nature, goals, and competencies. Ann Intern Med 133:59–69

Beauchamp TL, Childress JF (2013) Principles of biomedical ethics, 7th edn. Oxford University Press, New York

Chiavari L et al (2015) Difficult choices for young patients with cancer: the supportive role of decisional counselling. Support Care Cancer. Published on line: 11 April 2015. doi:10.1007/s00520-015-2726-5

Dauwerse L, Stolper M, Widdershoven G, Molewijk B (2014) Prevalence and characteristic of moral case deliberation in Dutch health care. Med Health Care Philos 17:365–375

Fiester A (2015) Neglected ends: clinical ethics consultation and the prospects for closure. AJOB 15(1):29–35

Gaucher N, Lantos J, Payot A (2013) How do national guidelines frame clinical ethics practice? A comparative analysis of guidelines from the US, the UK, Canada and France. Soc Sci Med 85:74–78

Jiwani B (2015) Ethically justified decisions. Healthc Manage Forum 28(2):86–89

Jonsen AR, Siegler M, Winslade WJ (2010) Clinical ethics: a practical approach to ethical decisions in clinical medicine, 7th edn. McGraw-Hill Medical, New York

Larcher V, Slowther A-M, Watson AR (2010) Core competences for clinical ethics committees. Clinical Ethics 10(1):30–33

Lawson AK et al (2015) Psychological counselling of female fertility preservation patients. J Psychosoc Oncol. Published online 21 May 2015. doi:10.1080/07347332.2015.1045677

Matchett NJ (2015) Philosophical counseling as an alternative process to bioethics mediation. AJOB 15(1):56–58

Molewijk B, Widdershoven G (2007) Conference reports: report of the Maastricht meeting of the European clinical ethics network. Clinical Ethics 2(1):42–45

Steinkamp N, Gordijn B (2003) Ethical case deliberation on the ward. A comparison of four methods. Med Health Care Phil 6:235–246

Stolper M, Molewijk B, Widdershoven G (2014) Learning by doing. Training health care professionals to become facilitator of moral case deliberation. HEC Forum. doi:10.1007/s10730-014-9251-7

Tapper EB (2013) Consults for conflict: the history of ethics consultation. Proc (Bayl Univ Med Cent) 26(4):417–422

Tarzian AJ (2013) Health care ethics consultation: an update on core competencies and emerging standards from the American Society for Bioethics and Humanities' core competencies update task force. AJOB 13(2):3–13

Tarzian AJ et al (2015) A code of ethics for health care ethics consultants: journey to the present and implications to the field. AJOB 15(5):38–51

Ethical Counselling for Patients

Giovanni Boniolo and Virginia Sanchini

Abstract In this chapter, we propose a methodology of Ethical Counselling addressed to patients and/or to their relatives. We show to what extent this is strongly grounded in a robust philosophical tradition: the Aristotelian practical philosophy. It is emphasised that such a methodology has been thought in order to help patients to make an aware ethical choice after having analysed what we call their *Personal Philosophy*, that is, their more or less systematic set of personal values, ideas and religious beliefs.

Keywords Ethical counselling · Patients · Method · Personal Philosophy · Aristotelian philosophy

Too many times philosophy has derogatorily been considered as a purely theoretical discipline without any impact on real life: mere words without any effect over the practice, very far from the empirical results found in biology, or the treatments offered in the clinics, or even the numbers coming out from physical theories. In contrast, the idea lying behind the proposal of an Ethical Counselling for patient's choices is precisely inspired by the opposite conviction, namely that philosophy is a life-changing enterprise having a relevant bearing over human relations and, in particular, over agency. This idea has a long-standing tradition within the history of philosophy, both in the continental and in the analytic tradition. To give some concrete examples, in 1984, Achenbach's *Philosophische Praxis* explicitly claimed that the time was ripe for philosophy to give an orientation to life. Some years

G. Boniolo (✉)
Dipartimento di Scienze Biomediche e Chirurgico Specialistiche, University of Ferrara, Ferrara, Italy
e-mail: giovanni.boniolo@unife.it

G. Boniolo
Institute for Advanced Study, Technische Universität München, Munich, Germany

V. Sanchini
Department of Experimental Oncology, European Institute of Oncology (IEO), Milan, Italy
e-mail: virginia.sanchini@ieo.eu

© The Author(s) 2016
G. Boniolo and V. Sanchini (eds.), *Ethical Counselling and Medical Decision-Making in the Era of Personalised Medicine*, SpringerBriefs on Ethical and Legal Issues in Biomedicine and Technology, DOI 10.1007/978-3-319-27690-8_2

before (1981), the French historian of philosophy Pierre Hadot, through his *Exercices spirituels et philosophie antique*, argued that philosophical theories from ancient Greece were actually "spiritual exercises", i.e. "techniques" to better cope with the troubles of life, thus qualifying as "practices [...] intended to effect a modification and a transformation in the subject who practices them". About a decade later in the USA, Martha Nussbaum published *The Therapy of Desire*, in which she defined ancient philosophers as "physicians", being they concerned with those pathologies affecting human life and decisions (see also Hamlyn 1992).

If we are willing to accept what Hadot, Achenbach, and Nussbaum (to name just a few) suggested, philosophy ceases to be considered as a mere theoretical enterprise and starts to acquire a pivotal role in orienteering our choices and life. This is precisely the line of though in which our approach of Ethical Counselling is grounded. Indeed, as we will argue, philosophical reasoning—as instantiated in the method we are proposing—can play a very important role in the improvement of human decision-making. This occurs in particular when some philosophical issues are involved, as it happens whenever clinical decisions intersect ethical questions.

By taking into consideration this point, in this chapter we propose a methodology grounded in the history of philosophy, which could be applied anytime clinical decision-making crosses with ethical decision-making. The type of subject towards which this methodology is directed is the patient who finds himself/herself in front of a difficult ethical decision or an ethical dilemma, and voluntarily asks for a consultation. We distinguish it from another methodology that we are going to present in the following chapter, which is addressed to clinicians. Notably, these two methodologies differ, not only because they are directed towards two different kinds of stakeholders, but also because their respective aims are different. On the one hand, the methodology presented in this chapter aims at helping patients to clarify their moral world view, thus helping them in making decisions that are consonant with their values and beliefs. On the other hand, the second methodological chapter aims to provide clinicians with a tool to support patient's decision-making on the ethical issues *he/she* is confronted and has to cope with.

1 The Aristotelian Practical Philosophy

Twenty years before Achenbach's *Philosophische Praxis*, another relevant philosophical event occurred in Germany. In 1960, H.G. Gadamer published his masterpiece *Wahrheit und Methode*, which marked the beginning of the so-called renaissance of the Aristotelian practical philosophy (see Riedel 1972–1974; Knight 2007). This work can be considered one of the first (if not the first) systematic attempt to show in which sense and by means of which methodology, philosophy could be a way of dealing with the troubles of life, in particular with the complicate decisions we are all confronted with (see Lobkowicz 1967).

This Aristotelian approach is also at the basis of our view of Ethical Counselling for Patients, this latter considered as a service aimed at helping them in solving their ethical dilemmas through the tools of philosophy (and, therefore, that philosophy has a practical validity in orienteering human decision-making).

Very briefly, the term "practical philosophy" appears for the first time in *Metaphysics* (II, 1, 993 b 19–23), where Aristotle distinguishes it from metaphysics, that is, the first philosophy, or theoretical philosophy. The latter has as its main purpose the investigation of truth, whereas the former deals with human action (*praxis*) in order to ameliorate it (*eupraxia*) and, thus, to improve human agency. Claiming that practical philosophy has to do with action means affirming that its domain regards decision-making. Indeed, what will be done in terms of actions will depend upon the agent's choices (*prohairesis*) (*Metaphysics*, VI, 1, 1025 b 22–24).

The place in which Aristotle better develops his account of practical philosophy is, however, the *Nicomachean Ethics*. At the end of *Book I*, he distinguishes between the intellectual and the ethical virtues. The former have to do with reason (*dianoia*), the latter (the moral excellences, *areté*) with characters, customs, and behaviours (*ethos*). In *Book VI*, he claims that philosophical wisdom (*sophia*) is the virtue of theoretical philosophy, while practical wisdom (*phronesis*) is the virtue of practical philosophy. By *practical wisdom* Aristotle means man's capacity of deliberating well, that is, the ability of deciding our life-goals and the most effective means to reach them.

The correlation between practical philosophy and its virtue, the practical wisdom, should now start appearing clearer. Practical philosophy consists in the *examination* of the different opinions that are on the stage in order to find out the best one/s through practical wisdom. Thus, in order to choose well, a man should, first and foremost, examine (*exetazein*) the epistemological plausibility and the logical validity of what is on the stage concerning a possible (good) action and, secondly, make the choice on the basis of what constitutes the best option for him.

A last point is necessary in order to complete the picture. A real good decision-maker is someone who necessarily possesses another ethical virtue: temperance (*sophrosyne*). This is the virtue that prevents passions from ruining a proper deliberation.[1]

Summing up, in order to bring a good decision, one should: (i) *control* his irrational part with the help of temperance; (ii) *examine* the situation from a rational standpoint, therefore evaluating the epistemological plausibility and the logical tenability of the pros and cons of each option; (iii) *deliberate* with the help of one's practical wisdom, in favour of what constitutes the best option for him/her.

These three conditions are of extreme importance for us, since we consider them as the peculiar features of our own account of Ethical Counselling for Patients. Indeed, whenever a patient finds himself/herself in the condition of taking an ethical

---

[1]To be philologically correct, in *Nicomachean Ethics* Aristotle uses the concept of *sophrosyne* in a narrower way in comparison with what we are saying in this chapter. The same term, however, was used in the sense we need by other Greek philosophers. For example, by Plato in *Cratylus*.

decision or solving an ethical dilemma concerning a diagnostic or therapeutic path, the ethical counsellor should exercise an advisory role and, in particular (i) help the patient to tame his/her emotions and (ii) examine, in the Aristotelian sense, his/her possible moral options and their consequences, and finally (iii) assist the patient to use his/her practical wisdom and thus to individuate what constitutes for him/her the best decision and course of action.

2   The Methodology

Ethical Counselling for Patients is therefore a service aimed at making explicit and *examining* (in the Aristotelian sense) the patient's Personal Philosophy in order to unlock his/her decisional paralysis when addressing clinical choices, imbued with ethical beliefs.

The Ethical Counselling for Patients is articulated in an ordered succession of four well-defined phases: (1) *relational phase*, (2) *medical assessment phase*, (3) *ethical analysis phase* (in turn, made up of three sub-steps), and (4) *wrap-up phase*. Each step will be presented in detail here, in relation to the Case 1 (presented in The Plan).

2.1   Relational Phase

The ethical counsellor starts creating a cooperative relationship by explaining the goals of this service and individuating the aim of the colloquium together with the patient. In particular, in this first phase the ethical counsellor presents the ethical counselling as a discretionary service at patient's disposal. This, in turn, means briefly explaining to the patient that ethical counselling focuses on the ethical dilemmas arising in the clinical setting. Moreover, the ethical counsellor tells the patient that the aim of this consultation is first of all to identify what is his/her ethical dilemma and to establish how to address it.

> *The ethical counsellor presents himself/herself to Giovanna and explains the basic elements of an ethical counselling service. In particular, she helps Giovanna understand to what extent the choice whether to test or not her two children is not only a clinical choice but involves also an ethical dimension. In other words, the ethical counsellor helps Giovanna understand what her ethical dilemma is. After having done this and asked Giovanna whether she would like continuing the consultation, the ethical counsellor establishes with her the specific aim of the colloquium; that is, what is the goal she would like to reach through it. In this case, the specific aim of the colloquium could be to better understand which advantages and disadvantages she had, should she immediately test the children or wait to test them until they reach the majority age. If these aspects are already clear to Giovanna, another aim could consist in investigating the moral reasons supporting the two different options, so as to let her have a clearer picture of the entire situation, thus supporting her downstream decision.*

2.2 Medical Assessment Phase

The ethical counsellor focuses on the patient's knowledge in order to verify if he/she is provided with the relevant medical information to make a properly informed choice. Whether the patient is uninformed, the ethical counsellor may suggest him/her consulting the physician to clarify any aspect left unclear. Obviously, the patient should not perceive this process as if someone is trying to verify her knowledge. By contrast, it is very important that the ethical counsellor sets up this process in a dialogic manner, by behaving as if he/she was unaware of the medical information and asked the patient to provide him/her with it.

> *The ethical counsellor tries to assess whether Giovanna has a clear understanding of the Li-Fraumeni syndrome and of its possible clinical consequences, of the peculiarity of a hereditary disease, of the significance of a genetic test, and of the probabilistic meaning of the outcomes. This is achieved by speaking with Giovanna about the aforementioned concepts, so as to be sure that her choice is grounded in valid medical information. In the case in which Giovanna was not provided with enough information–or she had an insufficient understanding of it–the ethical counsellor would ask the clinician to intervene to fill up those knowledge gaps.*

2.3 Ethical Analysis Phase

Once the examination of the medical aspects is concluded, the ethical counsellor supports the patient to analyse his/her ethical dilemma. This is done by helping him/her to investigate all the available options from an ethical standpoint, that is, by considering what are the values that each scenario promotes or inhibits and, therefore, what are the moral reasons in favour and against each clinical option in the light of the patient's Personal Philosophy. Fleshing out the values promoted or privileged by any option could help in gaining a different and more complete picture of the decisional scenario. This phase articulates in three different substeps.

2.3.1 Ethical Assessment Phase

The ethical counsellor explores patient's (implicit or explicit) ethical principles, assumptions, values, and beliefs in order understand and to help him/her in "unveiling" his/her Personal Philosophy. By doing so, both the patient and the counsellor may gain a deeper understanding of what counts for the former as morally valuable: a necessary step to cope with the options of the ethical dilemma.

> *The ethical counsellor asks Giovanna why this situation is dilemmatic for her, what are the values she believes ought to be served in this situation and whether there are some values that she sees at odds here. By exploring Giovanna's Personal Philosophy and the specific way she interprets her case, the ethical counsellor could, for example, realize that the mere*

idea of testing the two children worries Giovanna a lot because of the many cases of disease in her family that make her lean towards believing that the two children could be subjected to the same fate. Therefore, the hypothesis of postponing the test appears justified by the requirement to protect the children from the overall path that they should follow if a positive result were found. On the other hand, she could also think that if she really wants to protect them she should know in advance the possible diseases her children could be predisposed, and thus that the test should be performed. Moreover, by putting herself into her children's shoes, she might think that they would rather prefer to be informed. The ethical counsellor possesses now some elements to understand to what extent this case sounds dilemmatic to Giovanna. She ranks as her primary value the one of "protecting her children", and yet she recognizes that there are two contrasting ways in which she can specify such value in the present scenario. On the one hand, she could protect them from knowing a possibly sad truth; on the other hand, she could better protect them from the onset of future diseases only if she comes to know whether they are or not carrier of that specific mutation.

2.3.2 Ethical Comparative Phase

The ethical counsellor helps the patient in elaborating a ranking of values. In particular, the ethical counsellor asks him/her to provide his/her actual ranking of values, and to think of it in relation to the experience of illness; that is, whether and how the direct/indirect experience of illness has changed it.

Once the main elements of her Personal Philosophy have been pointed out, the ethical counsellor asks Giovanna to reflect upon her grounding values and then to provide a ranking of them. In this case scenario, the ethical counsellor could ask Giovanna to rank those principles and elements mentioned before such as protecting her children from knowing (implemented into the exercise of what the ethicists would call "the right not to know", on behalf of the two children), the duty of being informed in the case of personal safety, and so on. Moreover, the ethical counsellor could ask Giovanna whether she thinks that the way in which she ranked her values has changed after having been informed of the potential disease of the children.

2.3.3 Perspective Phase

The ethical counsellor asks the patient to focus again on the available medical options and to apply the values just discussed to the case, so as to see the relationship between the ethical values and the clinical options. By doing so, the patient should be able to provide a specific and personal weight to each medical option also from an ethical standpoint. Moreover, the ethical counsellor could ask him/her to make a though experiment and to imagine what would follow in terms of consequences for him/her and for his/her relatives from the adoption of one option over another.

As it has been already pointed out, Giovanna finds herself in the troubling dilemma of choosing between non-information and information within the context of protecting her children. Actually, what Giovanna realizes while discussing with the ethical counsellor is that these two elements, far from being incompatible, are strictly intertwined: indeed she understands that, to some extent, she has higher chances of protecting her children by informing them and, therefore, by testing them. Moreover, she realises that the real incompatibility is between pursuing an alleged right not to know in place of her two children, and pursuing a duty to inform the two children about such a possibility. The ethical counsellor, on the one hand, has helped Giovanna to put on the table the main elements of her Personal Philosophy triggered by this complex decision, while, on the other hand, assisting her to partially rethink them (unmasking false incompatibilities) and to apply them to the specific case so as to provide them with the awareness of what values and ethical options correspond to the available medical options. Moreover, each Giovanna's choice—to test or not to test the children, to tell or not to tell them the outcomes, when to tell them the outcomes—has a lot of consequences for her, her children, her husband and relatives. During this phase the ethical counsellor helps Giovanna in eliciting and clarifying each decisional path and the ethical and existential load each one is carrying.

2.4   Wrap-up Phase

The ethical counsellor summarises what has been found asking the patient whether he/she is satisfied or whether a phase of the counselling process needs to be rerun. It is important to note that even if the aim is to break a decisional paralysis, it is not necessary that the patient makes a decision by the end of the colloquium (or the series of colloquia). In other words, the important feature here is that he/she has clearer ideas regarding the options at stake, thus having gained all the relevant elements he/she might need in order to make a choice.

The ethical counsellor summarises what has been found asking Giovanna, whether she is satisfied or whether a phase has to be rerun.

Note that over the entire course of the meeting/s, the ethical counsellor pays attention to the patient's justificatory abilities as well as on possible reasoning fallacies and biases.

The ethical counsellor tries to investigate in a deeper way to what extent Giovanna's values and perspectives are properly justified and whether her position is logically consistent or not. This exercise does not have the aim of changing or ameliorating (in particular according to the counsellor's view) her way of seeing life, but to see whether some of Giovanna's doubts are grounded in logical mistakes rather than in real worries and/or, above all, in conflicts of values.

References

Aristotle (1998) The metaphysics. Penguin, London
Aristotle (2004) The Nicomachean ethics. Penguin, London
Gadamer HG (1960) Truth and method. Sheed and Ward, London 1989
Hadot P (1981) Philosophy as a way of life. Blackwell, Oxford 1995
Hamlyn D (1992) Being a philosopher: the history of a practice. Routledge, London

Knight K (2007) Aristotelian philosophy: ethics and politics from Aristotle to Macintyre. Polity Press, Cambridge

Lobkowicz N (1967) Theory and practice: history of a concept from Aristotle to Marx. University of Notre Dame Press, Notre Dame

Nussbaum M (1997) The therapy of desire: theory and practice in Hellenistic ethics. Princeton University Press, Princeton

Riedel M (1972–1974) Rehabilitierung der Praktischen Philosophie. Rombach Verlag, Freiburg

Ethical Counselling for Physicians

Giovanni Boniolo and Virginia Sanchini

Abstract In this chapter, we propose a methodology of Ethical Counselling addressed to the physician and/or the medical team. The aim of this second methodology is to help clinicians to have a more complete picture of the moral issues raised by the clinical case they encounter in their daily practice. We show that this methodology is philosophically grounded upon the way Medieval philosophers structured their argumentations.

Keywords Ethical Counselling · Clinicians · Problem analysis · *Status quaestionis* · Medieval *disputatio*

In the previous chapter, we have discussed and proposed a methodology of Ethical Counselling directed towards patients, aimed at supporting them when asked to address and solve ethical dilemmas arising from their clinical situations. However, the same methodology, when partially modified, appears to have some benefits not only for patients, but also for those figures indirectly affected by ethical dilemmas, i.e. physicians. Indeed, an extensive literature exists as to the importance of considering the physician not only as the professional figure who is expected to cure the patient, but also as the partner for both therapeutic and non-therapeutic decisions in care pathways.

Nevertheless, since physicians are not supposed to solve patient's dilemmas, the purpose of an Ethical Counselling service, when directed towards them, should be partially rethought. Indeed, if in the case of patients the aim of this service is to facilitate their awareness of their Personal Philosophies so as to assist them in their complex clinical decisions, when directed towards physicians Ethical Counselling

G. Boniolo (✉)
Dipartimento di Scienze Biomediche e Chirurgico Specialistiche, University of Ferrara, Ferrara, Italy
e-mail: giovanni.boniolo@unife.it

G. Boniolo
Institute for Advanced Study, Technische Universität München, Munich, Germany

V. Sanchini
Department of Experimental Oncology, European Institute of Oncology (IEO), Milano, Italy

G. Boniolo and V. Sanchini (eds.), *Ethical Counselling and Medical Decision-Making in the Era of Personalised Medicine*, SpringerBriefs on Ethical and Legal Issues in Biomedicine and Technology, DOI 10.1007/978-3-319-27690-8_3

is rather aimed at conceptually elucidating the whole scenario also from an ethical standpoint, so as to allow them to deal appropriately with patients' ethical questions. This does not mean that physicians are supposed to be provided with as much ethical knowledge as an ethicist is supposed to have. Rather, they should be provided with that ethical knowledge that enable them establishing, also very generally, the framework within which the case they are facing is ethically located.

This recognition implies that the Ethical Counselling for Physicians has to be structured differently from the one directed to patients. More specifically, we propose that Ethical Counselling for professionals should be rooted in the Medieval *disputatio* and, in particular, in its first part: the *status quaestionis*, that is, the presentation of the problem at issue and of its justified solution/s. In fact, the physician facing patient's ethical questions should not try to solve them but to be informed about the wider moral context in which it is framed and about its possible justified solutions. Put differently, being the patient's dilemma intrinsically connected to his/her Personal Philosophy, the physician is not entitled to decide on his/her behalf. By contrast, since physicians ought to help the patient deciding according to his/her Personal Philosophy, it could be useful for the former to understand the dilemma in its entirety and complexity with the guide of a well-constructed Ethical Counselling service.

1 The Medieval Origin

Universities, since their origin in the XII century,[1] were the institutional places in which knowledge was both produced and transmitted from mentors to disciples. They have been the cornerstone on which modern and contemporary sciences and humanities have been built and the places where their methodological characterization was standardized.

Although each Medieval university had its own regulation, there was some degree of homogeneity in the didactic structure, based on *lectiones* (lectures) and *disputationes* (disputes). Usually, there was a first level, that of the faculty of *artes liberales* (liberal arts), in which a scholar could be enrolled at the age of 14–15. Then he had the possibility of proceeding to a second level and choosing amongst Medicine, Law, and Theology.

The Faculty of Arts lasted around 4 years. In the first two, a student had *lectiones* on *Trivium* (logic, rhetoric, grammar) and on *Quadrivium* (geometry, arithmetic, astronomy, music). In the second year, he began the disputes (*disputationes*). At the end, he could obtain a baccalaureate (*Artium Baccalaureus*).

---

[1]As known, between the twelfth and the thirteenth centuries, the first universities arose (even before, if we consider the Medical School of Salerno: ninth–tenth centuries) as pearls of a precious necklace: Bologna (1088), Paris (1090), Oxford (1096), Montpellier (1150), Cambridge (1209), Padova (1222), Napoli (1224), Toulouse (1229), Siena (1240), Coimbra (1290), Rome (1303), Perugia (1308), Firenze (1321), Pisa (1343), Valladolid (1346), Prague (1348), Pavia (1361), Krakow (1364), Vienna (1365), Heidelberg (1386), Ferrara (1391), etc.

As the reader may observe, independently of whether a student wanted to become a physician, a lawyer, or a theologian, he had to extensively study and practice the *Trivium*. It was unthinkable that a physician, a lawyer, or a theologian was not able to reason (logic), to write (grammar), and to persuade (rhetoric) properly and correctly.

Even if each university had its own, more or less detailed, regulation concerning the didactic activities, the curriculum, as mentioned, was basically identical everywhere: above all lectures and disputes.

The *disputatio* was structured as a debate between two students of an equal level of education and chaired by a master. This latter introduced a topic (*casus*, or *thema*) and posed a question (*problema*, or *quaestio*). Then, one of the students, the *opponens* (or *impugnans*), had to offer some arguments to which the other student, the *respondens* (or *defendens*), had to reply with counter-arguments. At the end, the master stopped the dispute by proposing his view and his arguments, which were considered as the *determinatio* or *solutio*.

Therefore, the medieval student was first and foremost asked to expose the *status quaestionis*, by going through the following steps:

1. *outline the relevant context*, by means of which he/she located the problem.
2. *briefly state the problem to be addressed*. This was done in order to better focus on the issue to be addressed and to verify that what was under discussion was clearly understood.
3. *define the terms to be used*. In such a way, any possible source of terminological ambiguity or confusion was eliminated.
4. *show the relevance of the problem and the impact of its solution*. This was done to exclude that an irrelevant issue was under analysis.
5. *expose the solutions which are alternative to that one you want to support and argue why they are not acceptable*. This was done in order to highlight two aspects: (i) it is uselessly arrogant to offer a new solution without taking into account the solutions already proposed by others; (ii) it is pointless to propose a new solution, if the already existing ones have not proved to be incorrect or argumentatively flawed.
6. finally, *formulate the solution*.

After the presentation of the *status quaestionis*, the student had to argue in favour of his proposed solution.

To sum up, two steps were therefore considered essential for the proper examination of an issue: (1) exposition of the *status quaestionis* and (2) exposition of the justification of the proposed solution (on this method and on its pervasiveness and usefulness, see Boniolo 2012).

2 The Methodological Proposal

As already said, the Ethical Counselling when devised for physicians deals much more with a problem-conceptualizing rather than with a problem-solving dimension. This is due to the fact that the methodology hereby presented is primarily aimed at guiding physicians to have the most complete picture of the ethical dilemma they are confronting with, so as to enable them to deal, in a non paternalistic way, with those who are the real final actors of the decision: the patient and/or his/her relatives.

By taking all of this into consideration, the first step of a well-constructed *disputatio*—that is, the presentation of the *status quaestionis*—is exactly what meets our needs. This means that the Ethical Counselling for Physicians should move, in a series of meetings, along the following seven phases (illustrated by analysing Case 1, in "The Plan"):

1. *Presentation of the clinical case.* Here the physician is asked to present the case from a clinical standpoint, so as to have an overall medical picture of the problem to be addressed and of the possible clinical paths the patient might decide to follow.
   The physician briefly summarizes to the ethical counsellor the case of Giovanna as it has been reported above.
2. *Presentation of the consequences for the patient to choose in favour of one medical option over another.* Here the physician is asked to present what the clinical consequences of the possible medical options are. This is done in order to clarify the data and to ground following ethical analysis on qualified scientific information.
   The physician explains to the ethical counsellor that the mother could make three different choices from a medical standpoint:

   - *The mother could decide not to test their children at all. The consequence of this is that she is preventing her children from knowing some relevant aspects about their future, which could psychologically hurt them. Nevertheless, she could make an active surveillance by doing preventive medical exams, which otherwise would not be performed.*
   - *The mother could decide to test the children, even if they are incompetent beings. In this case the consequence is that she is forcing her children to know some relevant aspects concerning their future, which will have both some positive and negative aspects, as just shown above.*
   - *The mother could decide not to test the children until they become competent beings. By doing so, she is postponing the provision of information to a future time, which might or might not be too late in order to plan and accept a different life.*

   An additional observation seems relevant:

   - *In order to make the choice, the question whether a pharmacological pre-vention of the genetic disease exists could play a very important role in deciding whether to test the two children. On the other hand, one might make the opposite decision of not testing the children, by considering the huge number of medical tests the children will be obliged to undergo in case of positive results, which might trigger a refusal of further tests when they will get older and the risks correlated to this mutations will be higher.*

3. *Presentation of the ethical issues raised by the clinical case.* Here the ethical counsellor helps the physician to examine the ethical dilemma lying behind the case so as to map the case also from an ethical standpoint.

 This case places itself, from a bioethical standpoint, in the classic debate concerning genetic testing which sees in the trade-off between the so-called right not to know and the duty to inform its main ethical dilemma (Andorno 2004). These two concerns are indeed clearly competing in this case: we have a woman, who is also a mother, carrying a clinically significant genetic mutation, and she is aware of her genetic make up, whereas the two children are not. Moreover, being a germline mutation, the children might also present the same mutation, with 50 % probability of having inherited it from her. Given this picture, what is the more legitimate option to pursue? Should the two children be informed about their genetic make up, or should they be allowed to exercise their right not to know? This situation seems to be further worsened if we consider that, being children considered as paradigmatic examples of incompetent agents, they cannot autonomously decide by themselves. Rather, it is the mother who should choose for them. Therefore, this case study presents the following ethical problematic situation: to what extent are these children entitled to the so-called right not to know? Or, alternatively, do we have any duty to inform them about their possible genetic susceptibility to cancer? Moreover, what kind of import has the fact that children are incompetent agents on the decision to be made (Robertson, Savulescu 2001)?

4. *Definition of the ethical terms.* Before analysing the case from an ethical standpoint, the ethical counsellor properly defines the main terms to be used. This step is essential in order to avoid any possible source of semantic ambiguity or lexical confusion.

 The ethical analysis of this case is completely focused on two main terms:

- *The right not to know. This right is often related to genetic information and might be translated in the right not to know one's genetic make up. The right not to know is usually justified by the risk of serious psychological harm, where there is no symmetric correspondence between the information disclosed and the concrete therapeutic options that might then follow, or where it appears too burdensome, both for the patient and the relatives, to hold such information.*

- *The duty to inform. Also in this case, we are referring to the duty to inform the individual of his/her genetic make up. This principle places the limits of not communicating to the individual his/her genetic information, when such information deals with the germline and, therefore, may be in common with his/her relatives. When it deals with shared genetic information, the right not to know cannot be considered as an absolute right at all but should be balanced with a duty to inform those to whom the information may concern.*

5. *Presentation of the ethical arguments in favour and against each medical option the patient might decide to follow.* Here the ethical counsellor, in a dialogue with the physician, puts on the table all the ethical pros and cons of each medical alternative the patient might decide to pursue. This is done in order to allow the clinician to clarify all the possible ethical viewpoints present in the specific case, and allow him/her to make a preliminary trade-off between both ethical values and medical options of the case under consideration.

   In favour of the decision of not testing the children at all, we could first of all present a non-maleficence-based argument, according to which if no readily medical benefit is available for the condition at issue, there is a serious risk that this information might psychologically harm the children. Secondly, we could support the same position starting from an autonomy-based argument, whereby informing the children about this medically relevant information breaches their future right to decide as autonomous adults about whether to keep this information confidential. In favour of the decision to test the children, we might claim that being incompetent agents, children should be subjected to these tests only for the sake of their best interest. Even if strong disagreement exists about how to define and enforce a meaningful conception of "best interest" in this situation, those who support this position claim that this information may be significant for the development of the child's autonomous life plan. Understanding some genetic predispositions, they argue, is an important piece of individual self-knowledge, partaking in the development of one's conception of a good life. If knowledge regarding risk of developing some diseases influences decisions that make our lives full (e.g. reproductive choices of young women who are found to be positive to risk-enhancing BRCA1/2 mutations), then why should we withhold this information from children? This knowledge is as important to them, as it is to any other agent.

6. *Examination of the patient's beliefs.* Here the counsellor helps the physician to understand whether the patient has ideological, philosophical, or religious beliefs that could influence his/her decision-making process and shift the balance in favour of one option over another.

   The ethical counsellor tries to investigate whether there are some elements belonging to Giovanna's spiritual/personal/religious beliefs that could influence her decision in a preferential way. Although the best way to do this would be to directly interview Giovanna, since the purposes of this consultation are more explorative than decisional (the aim is to have a clarification of the ethical issues raised by the case, not to decide on the behalf of Giovanna), this analysis is performed starting from the knowledge directly or indirectly available to the physician.

7. *Conclusive summary.* In this last step, the ethical counsellor proposes an ethical wrap-up of the case in question to the clinician, so as to be sure that he/she has a complete ethical picture of the case.

References

Backes CH, Moorehead P, Nelin LD (2011) Cancer in pregnancy: fetal and neonatal outcomes. Clin Obstet Gynecol 54:574–590

Boniolo G (2012) The art of deliberating. Democracy, deliberation and the life sciences between history and theory. Springer, Heidelberg

Molho RB et al (2008) The complexity of management of pregnancy-associated malignant soft tissue and bone tumors. Gynecol Obstet Invest 65:89–95

Peccatori FA et al (2013) Cancer, pregnancy and fertility: ESMO clinical practice guidelines for diagnosis, treatment and follow-up. Ann Oncol 24: vi160–vi170

Nocebo and the Patient–Physician Communication

Luana Colloca and Yvonne Nestoriuc

Abstract This chapter tackles one of the most delicate issues of the Ethical Counselling: communication. Communication is the central skill required for a proper patient–clinician relationship, but it is also vital whenever a service of ethical counselling is implemented. We do not dwell upon how it should be made. Differently, we discuss its neurocognitive counterpart. That is, we show how a wrong way of informing patients about their clinical situation or about the ethical dilemma they are facing could induce dangerous nocebo effects.

Keywords Communication · Neurocognition · Neurobiology · Nocebo effects

For decades, nocebo effects—the adverse events related to negative expectations and anticipations—have been dismissed as purely psychological effects. Current research illustrates that nocebos produce behavioural, functional and body changes, a finding that may transform how patient–doctor communication is framed and practised. Indeed, providers' behaviours, environmental cues, the appearance of medical devices and mainly verbal communication can induce negative expectations that dramatically influence clinical outcomes in a variety of patient populations. Expectations can be created through anticipations of worsening via verbal suggestions and/or prior exposure to negative symptoms (for a review, see Colloca and Miller 2011).

L. Colloca (✉)
Department of Anesthesiology, School of Medicine, University of Maryland,
Baltimore, MD, USA
e-mail: colloca@son.umaryland.edu

Y. Nestoriuc
Institute of Psychology, Clinical Psychology and Psychotherapy,
University of Hamburg, Hamburg, Germany
e-mail: yvonne.nestoriuc@uni-hamburg.de

© The Author(s) 2016   29
G. Boniolo and V. Sanchini (eds.), *Ethical Counselling and Medical Decision-Making in the Era of Personalised Medicine*, SpringerBriefs on Ethical and Legal Issues in Biomedicine and Technology, DOI 10.1007/978-3-319-27690-8_4

Nocebo effects can significantly increase non-specific symptoms and complaints in patient populations, resulting in psychological distress, medication non-adherence and need for additional drug prescriptions to treat subsequent nocebo-induced adverse effects. For example, headaches, which are a common side effect of antidepressants, can result simply from the mention of headaches in the informed consent process as a potential side effect.

Here, we describe selected informative studies in the field of the nocebo phenomenon indicating how clinical outcomes are suitable for cognitive influences deriving from information and decision-making process. Needless to remark, this is a central topic for any ethical counselling service, being it based on a dialogue and crossing it the clinical communication. That is, any ethical counsellor should be aware of the negative impact that a wrong way of interacting with patient could have. By taking this into account, we conclude by providing an ethical framework that can guide any counsellor and healthy provider during the counselling process.

1 Relation Between Clinical Communication and Nocebo Effects

Some recent studies emphasize the connection between reported side effects in the placebo arms and the known side effects of particular drugs indicating genuine nocebo effects derived from the act of exchanging information in counsel with clinicians before deciding the therapeutic plan.

Caution is, however, needed. It is important to clarify that these effects can represent either an apparent or a true nocebo effect. For example, when patients report headache, it is likely that the adverse event represents merely an apparent nocebo effect. Hence, the side effects observed in the placebo group may reflect misattribution, reattribution as well as the natural history of the condition or common symptoms that everyone experiences, rather than true nocebo effects. From a methodological perspective, nocebo effects can be assessed either by including a no-treatment group that does not receive placebos or by including a group that is not informed about the side effects related to a treatment under investigation. These alternatives might present ethical constraints because of the intentional concealment of the information and therefore a potential violation of the patient's rights (Colloca and Miller 2011).

Communication of adverse effects often leads to withdrawal from the study. The action of mentioning gastrointestinal side effects during the consent process in a randomized, double-blind, placebo-controlled trial examining the benefit of either aspirin or sulfinpyrazone, or both drugs, for unstable angina pectoris influenced their occurrence. The authors found that the inclusion of potential gastrointestinal side effects in the informed consent forms resulted in a sixfold increase in gastrointestinal symptoms with consequent patient-initiated cessation of therapy (Myers et al. 1987).

Nocebo effects produced discontinuation and lack of adherence also in RCTs for statin drugs in population-based studies. In statin trials performed from 1994 to 2003, placebo groups presented a variety of symptoms such as headache (0.2–2.7 %) and abdominal pain (0.9–3.9 %). The general population presented even higher adverse event rates than those found in clinical trials of statin drugs, further indicating that information about adverse events can elicit symptoms that reflect the anticipated effects like a self-fulfilling prophecy (Rief et al. 2006).

2 Mitigating Nocebo Effects Through Optimization of Patient–Clinician Communication

Awareness of drug-related adverse side effects induces nocebo effects that can last as long as 6–12 months. A recent study suggests that the effects of the patient–clinician communication are not limited to motivate patients to adhere to a recommended treatment regimen, to choose a healthier lifestyle and to adopt better psychological attitudes, but also to avoid occurrence of nocebo effects. Patients diagnosed with benign prostatic hyperplasia (BPH) were treated with finasteride (5 mg). The treatment was described as a "compound of proven efficacy for the treatment of BPH", and then, patients were randomized to two distinct disclosure contents. One group was informed that the medication "may cause erectile dysfunction, decreased libido and problems of ejaculation, but these are uncommon"; the other group was not told at all about these side effects. The 6- and 12-month follow-up indicated that finasteride administration produced a significantly different rate of reported sexual side effects. Those patients who were informed about the possibility of sexual dysfunction presented higher rate of sexual adverse events (43.6 %) as compared to those in whom the same information was omitted (15.3 %) (Mondaini et al. 2007). Although concealment of adverse events is problematic in daily clinical practice and this approach is not ethically justifiable, these findings clearly show how content of the informed consent and the patient–physician communication impact the occurrence of certain adverse events raising the need to consider ethical acceptable approaches.

Recent advance in expectancy and nocebo research outlines the need to reconsider the importance of the patient–physician communication, adverse events induced by negative experiences and expectations in clinical counselling, as well as the need to incorporate framing strategies to maximize placebo effects and minimize nocebo effects. It is becoming evident that verbal instructions that are a crucial component of any counselling processes are powerful in triggering negative expectations with an impact on clinical outcomes. Therefore, a first step is the realization that clinician's words and attitudes can potentially facilitate or worsen symptoms' improvement and healing processes.

A potential solution to mitigate nocebo-related adverse events is based on framing strategies. The same information can be provided in a way to minimize distress, anxiety, fear and tension. A recent study in women at term gestation requesting epidural analgesia emphasized the need for careful and thoughtful disclosures by randomizing patients to either a commonly used description of the pain experienced during the epidural procedure or a soft version incorporating some positive elements, while the women were informed about the procedure (Varelmann et al. 2010). In performing the epidural analgesic puncture, the authors used two disclosures: (1) "You are going to feel a big bee sting; this is the worst part of the procedure"; and (2) "We are going to give you a local anaesthetic that will numb the area and you will be comfortable during the procedure". Likewise, we all agree that the former disclaimer described above reflects a standard way to communicate the effect of the procedural intervention, while the second description describes the procedure anticipating the benefit of the anaesthetic medication. Pain experience was assessed afterwards in a blinded fashion. Those women in labour who were told to expect pain like a bee sting during the local anaesthetic injection (nocebo group) reported significantly higher perceived pain as compared to those receiving the procedure along with gentle and positive words (Varelmann et al. 2010). This study emphasized how small changes in the way in which information is framed can strongly impact clinical outcomes, suggesting that it is possible and ethically acceptable to present information in a way that reduces nocebo effects while preserves patients' rights to be informed. These studies strongly encourage any counsellor to include in her/his practice elements such as an empathic communication that, while understands another person's experience from that person's perspective, is consistent with sincerity and professional integrity. The capacity to work with the counselee's concerns without being personally diminished and to show appropriate esteem to others and their understanding of themselves along with an effective deployment of skills and knowledge can play an important role in minimizing nocebo effects even when critical contexts may require a strong capacity to act in spite of known fears, risks and uncertainty.

3 Clinical Communication and Patients' Expectations

Patient–physician communication and framing effects become extremely important in medicine and oncology. For example, recent research by Nestoriuc et al. shows that it is feasible to promote beneficial placebo effects and minimize nocebo reactions in breast cancer women (Nestoriuc et al. 2015).

  Nocebo effects are of great relevance in oncology, since they add to the burden of symptoms in long-term survivors and often motivate patients to discontinue effective adjuvant treatments. Adjuvant endocrine therapy for breast cancer is the most widely prescribed cancer treatment in the world. Yet, non-adherence rates, which are mostly determined by adverse side effects, are substantial and lead to decreased survival. In a two-year study of a clinical cohort of 111 patients treated with hormone receptor-positive breast cancer, it was shown that long-term side effects and non-adherence were largely determined by patient expectations as opposed to medical factors such as type of endocrine treatment, or cancer staging. Holding negative expectations about endocrine therapy before starting the treatment increased the risk of nocebo-related side effects, low quality of life and non-adherence over two years of treatment. Expectation effects remained significant after controlling for multiple medical and psychological confounds (Nestoriuc et al. 2015). Expectations are potentially modifiable and thus can become a therapeutic target to prevent nocebo effects and related non-adherence. In the sample of women described above, nocebo-related effects accounted for relative risk reduction of 45 % (Nestoriuc et al. 2015). In other words, women holding positive or low negative expectations before the treatment experienced almost half the side effects than women with negative expectations. Thus, manipulating individual expectations might be a promising strategy to improve side effect burden, quality of life and adherence during longer-term drug intake.

Overall, patient beliefs and expectations about medicines are shaped by prior experiences and information provided by healthcare professionals. Medical informed consent procedures have great potential to modify patient expectations in order to optimize treatment outcome by maximizing placebo and minimizing nocebo effects. However, today's informed consent procedures are largely viewed as ethically and legally obligatory procedures to provide standardized patient information and prevent litigation. In clinical routine, they are mainly focused on potential risks and side effects of a given treatment within a one-size-fits-all approach. The downsides of this approach have been critically discussed, particularly regarding the harm it may cause by shaping negative expectations and triggering nocebo-related side effects in routine health care (Colloca and Miller 2011).

Clinical studies investigating the impact of modulated informed consent procedures are rare. So far, most of the empirical work has been conducted regarding risk and benefit communication, with special emphasis on the effect of different framing strategies on shared decision-making and illness behaviours. The results indicate that framing information might be an effective approach to improve patient outcome. Nevertheless, several factors have to be considered when studying framing effects. First of all, benefits and risks can be communicated verbally or numerically. There has been agreement that risk and benefit communication using numbers leads to better comprehension and higher patient satisfaction than messages using verbal descriptors. For example, numerical descriptors lead to a smaller overestimation of probabilities than words. Numerical information can be provided using absolute and relative risks. Converging evidence has shown that information framed as relative risks appears more convincing and, in the clinical context, may facilitate adherence better than information framed in absolute risks or numbers needed to treat. Second, risk or benefit information can be given gain-framed or loss-framed. This is done by focusing on the respective benefits or harms that can be attained, for example, by adhering or not adhering to suggested health behaviour. According to this classification, negatively framed side effect information is focusing on the number of patients experiencing a given risk (e.g. 38 out of 100 women taking tamoxifen will experience hot flushes), while positively framed information focuses on the number of people not experiencing the side effect (e.g. 62 out of 100 women will be free of hot flushes). Although it seems intuitively plausible that the later wording might lead to a less negative risk perception, the negatively framed wording remains today's standard in informed consent procedures. Overall, there is considerable disagreement as to when gain-framed or loss-framed messages (i.e. fear appeals) are most effective. Depending on the type of message to be framed, different moderating factors including the recipients' disposition and prior beliefs such as involvement with the issue, ambivalence, perceived benefit or riskiness, processing depth or informational coping styles need to be considered.

In a cluster-randomized trial of 44 Danish general practitioners, the effect of two quantitative formats of risk communication (i.e. prolongation of life versus absolute risk reduction), regarding the redemption of statin prescriptions, was studied (Harmsen et al. 2014). The gain-framed prolongation of life format resulted in significantly fewer patients accepting the cholesterol-lowering medication (5.4 %) than the loss-framed condition (25 %). One potential conclusion is that the gain in life expectancy was judged insufficient to warrant lifelong therapy. The authors discuss that patients faced with own health risks are very sensitive to different information formats. In addition to the level of effectiveness, patients' perceptions thereof might be a relevant factor influencing informed decision-making.

A third domain regarding message framing is the informational context in which informed consent is given (i.e. contextual framing). Information about possible adverse effects might be framed within the desired effects of the treatment with or without explicit information about the mode of action or mechanism involved in achieving these desired effects. The effects of this contextual framing have recently been experimentally evaluated by Heisig and colleagues (2015). Information about two adjuvant breast cancer treatments (i.e. endocrine treatment and chemotherapy) was given to 124 healthy women around the age of 50. Women were asked to imagine a scenario in which they had been diagnosed with breast cancer and were to receive adjuvant treatment. They were then randomized to one of four informed consent procedures, encompassing either positive or neutral contextual framing within an either personalized or standardized interaction with the healthcare provider. To comply with ethical requirements and to make risk information as comprehensible as possible, all groups received written numerical data on the most common and most serious potential side effects using natural frequencies. In the positively framed condition, explicit numerical information on the desired effects was added (i.e. reduction of cancer recurrence and metastases) and framed by explaining the mode of action (i.e. how the antioestrogen would help the body to withdraw growth factors from the tumour cells, waste away potential metastases and protect itself against new incidences). Women in the neutrally framed condition were not informed about expected benefits. They were simply told that the endocrine treatment would help lower their oestrogen levels. All information was provided in written form and handed over by a trained healthcare professional who either interacted rather business-like and referred questions to the oncologists in the standardized condition, or provided individualized feedback to participants' questions in the individualized condition. The informed consent procedure in the neutral framing condition with the standardized interaction was designed to resemble clinical routine. Effects of the modulated consent procedures on treatment expectations were analysed. The contextualized framing resulted in lower expectations of side effects, a more functional necessity-concerned balance (i.e. higher perceived necessity in addition to fewer concerns about the treatment), lower decisional conflicts and a higher adherence intention regarding endocrine treatment. A significant interaction showed that the condition resembling informed consent in clinical practice resulted in the least functional expectations. In the chemotherapy sample, effects were less pronounced. Significantly optimized treatment expectations were shown for necessity-concerned balance and adherence intention but not for side effect expectations or decisional conflicts. The results showed that optimizing treatment expectations through modified informed consent procedures is clinically feasible and effective (Heisig et al. 2015). Given the relevance of treatment expectations for nocebo effects and long-term clinical outcome in breast cancer patients, the documented effects of contextual framing are of clinical relevance and should be investigated in clinical samples. Potential influence factors that might help explain the divergent results for the two adjuvant treatments include differences in prior knowledge and benefit expectations between endocrine treatment and chemotherapy. Specifically, disclosing the relatively small expected benefits of chemotherapy for breast cancer (i.e. 20–30 of 100 potential cancer recurrences are prevented through chemotherapy) might have contributed to overall more negative treatment expectations and thus fewer effects of benefit framing. Benefit framing might be an effective strategy to prevent nocebo side effects. The effects need to be studies in conjunction with the described three and potential further domains of message framing.

4 Ethical Considerations

Information provided along with the administration of any active treatment is akin to walking a tightrope of communication. The goal standard should be to avoid untenable nocebo effects. However, concealment of information is debatable, as some patients may not agree to undergo the treatment because of their preferences and values. Nevertheless, healthy providers have an obligation to convey truthful information to patients.

In circumstances in which the patient is not exposed to serious risks, healthy providers can consider the "authorized concealment" approach in which patients might consent to receive information only about potential serious or irreversible harm. The consent process should inform the patient about the concealment and encourage her to report any experienced adverse event promptly. A potential alternative to the "authorized concealment" approach is conveying information by taking advantage of framing strategies and manipulating expectations in a positive manner.

A variety of studies have investigated the effects of framing information regarding risks and benefits of interventions on patient decision-making. For instance, a physician who is recommending a drug to a patient may communicate the proportion of patients who experience the side effects. Side effects, such as headaches or nausea, may be mentioned merely as a slight possibility. There is also a choice in communicating the probability of experiencing adverse effects based on extant research, either qualitatively or quantitatively. Furthermore, this information can be conveyed by focusing on the minority of patients who experience a particular side effect or by focusing on the majority of patients who do not experience the side effect. These different ways of framing side effect information can have differential impact on patients' perception of adverse events and, potentially, occurrence of nocebo effects. Further research is needed to explore the link between perceived risks and benefits of interventions and nocebo effects.

5   Future Directions

In conclusion, nocebo research provides evidence supporting the claim that the patient–physician communication has tremendous effects on shaping clinical outcomes, adherence to treatments and coping strategies. Both clinical and ethical counselling can be optimized to manipulate expectations positively and favour cognitive appraisals that influence treatment outcomes.

In the future, it is vital to educate providers and patients about nocebo research and its clinical implications (Colloca and Finniss 2012). This perspective is still poorly explored, but therapists, doctors and nurses should be encouraged to systematically tell the patient that some adverse effects occur as a result of informing them about the realm of side effects. This perspective would require an effort to educate both providers and patients about the mechanisms of nocebo effects and translate what we have learned in laboratory settings into daily counselling and practice. Health practitioners should also consider nocebo reactions and the link between conveying information and occurrence of certain adverse events. Importantly, concerns about trustfulness should not impede helpfulness and pragmatism that are two key morally relevant aspects guiding clinical practice, as well as any therapeutic decision-making process. Remarkably, all these considerations should necessarily be taken into account whenever an ethical counselling is provided. In particular, it is relevant in the case in which the ethical counselling is provided in service of patients and their decisions when they are informed by the clinician about possible diagnostic or therapeutic choices. But the proposed ethical framework should be considered even in the cases in which the ethical counsellor informs clinicians encouraging them to stay alert to the adverse events and side effects their patients are experiencing since some of them are nocebo effects occurring as a result of the ways in which they communicate with patients.

References

Colloca L, Finniss D (2012) Nocebo effects, patient–clinician communication, and therapeutic outcomes. JAMA 307(6):567–568

Colloca L, Miller FG (2011) The nocebo effect and its relevance for clinical practice. Psychosom Med 73:598–603

Harmsen CG et al (2014) Communicating risk using absolute risk reduction or prolongation of life formats: cluster-randomised trial in general practice. Br J Gen Pract 64(621):e199–e207

Heisig SR et al (2015) Informing women with breast cancer about endocrine therapy: effects on knowledge and adherence. Psychooncology 24(2):130–137

Mondaini N et al (2007) Finasteride 5 mg and sexual side effects: how many of these are related to a nocebo phenomenon? J Sex Med 4(6):1708–1712

Myers MG, Cairns JA, Singer J (1987) The consent form as a possible cause of side effects. Clin Pharmacol Ther 42(3):250–253

Nestoriuc Y et al (2015) Expectation-effects in endocrine therapy for breast cancer: a two-year prospective clinical cohort study (Under review)

Rief W, Avorn J, Barsky AJ (2006) Medication-attributed adverse effects in placebo groups: implications for assessment of adverse effects. Arch Intern Med 166(2):155–160

Varelmann D et al (2010) Nocebo-induced hyperalgesia during local anesthetic injection. Anesth Analg 110(3):868–870

Reasons and Emotions

Marco Annoni

Abstract This chapter outlines the importance of reasons-giving in, and for, Ethical Counselling. By *reason-giving*, we refer to that particular dialogical process whereby an agent comes to identify, articulate and appraise the various moral reasons structuring an ethical dilemma. In particular, within this chapter, we describe (i) the role of emotions with respect to reason-giving; (ii) why we should rely on reason-giving; (iii) how reason-giving enables to make explicit patients' Personal Philosophy; and (iv) why reason-giving is always an open-ended process.

Keywords Ethical Counselling · Dialogue · Reason-giving · Emotions · Personal Philosophy

Ethical dilemmas occur whenever an agent is seemingly forced to make a choice between two mutually exclusive courses of action neither of which appears to be perfectly morally acceptable. Thus, in the case of Giovanna already presented, she is facing an ethical dilemma because she is forced to decide between two incompatible courses of action–i.e. to test or not to test her children for a certain genetic mutation–, and yet both options seem to be wrong because both might harm them. A common trait of ethical dilemmas is that the agent who must decide is also uncertain about what she ought to do. This uncertainty arises because there seem to be some "good reasons" both to act and not to act in a certain way, and it is thus unclear how one should eventually behave.

Henceforth we shall use the general term "reason" as a shortcut to refer to any "justification" or "rationale" that can be used to defend or criticize a given course of action. In reference to ethical dilemmas or, more generally, to ethically problematic situations, we shall thus speak of "moral reasons", as the arguments or justifications

M. Annoni (✉)
National Institute of Biomedical Technologies (ITB), Milan, Italy
e-mail: marco.annoni.guide@gmail.com

M. Annoni
National Research Council (CNR), Milan, Italy

M. Annoni
Fondazione Umberto Veronesi, Milano, Italy

G. Boniolo and V. Sanchini (eds.), *Ethical Counselling and Medical Decision-Making in the Era of Personalised Medicine*, SpringerBriefs on Ethical and Legal Issues in Biomedicine and Technology, DOI 10.1007/978-3-319-27690-8_5

at stake always concern what agents should or should not do in certain circumstances. In this view, an ethical dilemma is therefore essentially a conflict between two competing series of moral reasons. Therefore, in order to understand and possibly resolve an ethical dilemma, one should first identify and then analyse the moral reasons from which it originates, assessing their respective force. For the sake of simplicity and precision, we shall henceforth call *reason-giving* the process whereby an agent comes to identify, articulate and appraise the various moral reasons structuring an ethical dilemma. Likewise, we shall call *moral deliberation* the part of any reason-giving process in which an agent finally "makes up her mind", i.e. forms a considered opinion on the ethical issue at stake.

In Ethical Counselling, reason-giving takes the form of an interpersonal dialogue between the counsellor and the counselee. As such, reason-giving lies at the very core of our take on it. In fact, the methodologies proposed for patients and clinicians represent two ways of guiding someone in the proper identification, articulation and assessment of the moral reasons structuring a given ethical dilemma. Both patients and clinicians engaging in the Ethical Counselling are thus also engaging in reason-giving, even if they follow two different methodologies because they pursue two different aims. For patients, the aim is to overcome a state of "decisional paralysis", eventually deciding how to cope with an ethical dilemma in the light of their Personal Philosophy. For clinicians, instead, the aim is not to decide, but to acquire the ability to identify, explore and further articulate the various moral reasons from which ethical dilemmas originate.

In this chapter, we shall explain why we consider reason-giving as an essential component for moral deliberation, and thus as one of the qualifying features of Ethical Counselling. In particular, we shall consider (i) the role of emotions with respect to reason-giving; (ii) why we should rely on reason-giving as we face ethical dilemmas; (iii) how reason-giving enables to make explicit patients' Personal Philosophies; and (iv) why reason-giving is always an open-ended process and might be useful even if it does not lead to solve the ethical dilemma at stake.

1   Emotions and *Reason*-Giving

Despite the centrality that our approach assigns to reasons, it is important to stress that we do not consider reason-giving as a cold, emotionless activity. On the contrary, we acknowledge that emotions and feelings play an essential role in our moral life and thus in Ethical Counselling. We do claim, however, that emotions provide us with a necessary but not sufficient element for reaching optimal moral deliberation, and thus practical wisdom. In order to unpack this claim, in this section we shall briefly explore the role of emotions with respect to reason-giving.

To begin with, it is useful to look at some recent empirical findings in the fields of cognitive sciences and experimental moral psychology. Traditionally, reason and emotions have been portrayed as opposite forces. Overwhelming empirical evidence, however, now suggests that emotions and deliberate thinking may instead be closely intertwined one with the other, and even inseparable. Emotions affect the way in which we perceive the world, directing our attention and perception by privileging emotionally relevant stimuli (Phelps et al. 2006). They shape our memory, as memories of emotional events have an increased vividness and persistence (Phelps 2004). They influence our decision-making, as the anticipation of the emotional consequences of our actions is one of the factors affecting our own choices (Brosch et al. 2013). More importantly, emotions seem to play a crucial role in moral judgment (Prinz 2006). Witnessing immorality often elicits negative emotions such as contempt, anger and disgust, while witnessing moral virtue often elicits positive ones (Avramova and Inbar 2013). Neuroimaging studies demonstrate that emotional structures are normally recruited in making moral judgments (Haubner et al. 2009). Lastly, it has been hypothesized that emotions may also lead to moralization, as for example when disgust is cultivated in order to foster a sociopolitical agenda aimed at excluding a group of people from a community (Wheatley and Haidt 2005).

Experimental moral psychology is a relatively recent field of inquiry, and therefore, more research is needed to further elucidate both the role of emotions with respect to moral judgments and the role of moral judgments with respect to human cognition in general. Nevertheless, empirical studies show that emotions and moral judgment are always closely intertwined. Consequently, no interpersonal process of reason-giving can ever be purely emotionless. This is especially true in the case of ethical counselling, a practice devised to aid people facing complex, emotionally loaded choices. Therefore, engaging in reason-giving does not demand that the consultant or the counselee adopt a cold, emotionless attitude. On the contrary, establishing a positive relationship between the counsellor and the counselee is key to ensure that a productive reason-giving process can properly occur. In particular, if the counselee is a patient facing a very difficult choice, mobilizing positive emotions such as trust or relaxation is functional to the goal of counselling.

Yet, reason-giving also requires the management of those emotions that would prevent it from taking place. An example would readily explain this point. Consider the case of a person who must take a blood test. Unfortunately, this person has always been scared of syringes. As she arrives at the hospital, she becomes suddenly anxious and she feels the sudden urge to leave and flee. In order for this person to opt for the blood test, at least two deliberate *efforts* are required. The first effort is the one needed to "pause" and resist the urge of immediately running away in fear without taking the test. The second effort is the one needed to calmly reflect on the various options at hand, reach a considered opinion about their pros and cons, and then act accordingly. Without these *acts of deliberate self-control* to constrain and eventually override our first automatic emotional responses, it would be impossible for this person to engage in reason-giving in this and similar cases. Thus, while emotions are an integral part of any reason-giving process, sometimes they should also be controlled or "tamed" because otherwise they would impair our ability to reason.

This point can be further elucidated by drawing a parallel with the teachings of Stoicism, one of the most influential schools of thought of classical antiquity. The Stoics distinguished between "emotions" and "passions" (*pathê*), using the latter term to denote all kinds of strong, automatic and uncontrolled emotional disturbances.[1] They believed that passions were at the root of all human suffering because they "enslaved" those experiencing them, impairing their capacity to reason and thereby their possibility of becoming truly free and happy. By contrast, the Stoic sage was perfectly free and happy precisely because he knew how to enjoy all positive emotions and how to overcome all the irrational fears and desires lying at the core of unhealthy passions. In stressing the importance of taming one's emotions as a necessary precondition for optimal reason-giving, our approach parallels —although on a smaller scale—what the Stoics preached about the perils of falling prey to one's unhealthy and uncontrolled passions. As the Stoic sage had to overcome all passions in order to achieve true rational happiness (*eudaimonia*), so those engaging in ethical counselling must strive to control their strong emotional responses in order to properly partake in fruitful courses of shared reason-giving with their consultants.

2 Why Reason-Giving Is Important in Dealing with Ethical Dilemmas

Why should we rely on reason-giving rather than just on emotions or "intuitions" as we face ethical dilemmas? The answer to this question depends on the context.

For clinicians, reason-giving is important because to say that one has "understood" an ethical dilemma is tantamount to say that one has acquired at least the capacity to (i) identify the conflicting moral reasons structuring such dilemma; (ii) define the key terms on which these reasons depend (e.g. "autonomy" and "harm"); (iii) appraise the relative force of each set of competing reasons; and (iv) indicate how the balancing of the two competing set of moral reasons would change in different hypothetical scenarios. Therefore, for clinicians, engaging in reason-giving is necessary for gaining a better understanding of the ethical issues at stake.

For patients, instead, reason-giving is the best methodology to ensure that what they eventually decide conforms to their Personal Philosophy, and thus with the goals and values they would deliberately choose. To clarify this point, recall the example of the person who must take a blood draw but is afraid of syringes. Refusing a blood-draw out of the fear may contrast with her goal of testing for

---

[1]According to the Stoics, the need to control one's passions as to allow some space for critical thinking was not limited to negative emotions (e.g., fear, sadness), but it extended to all cases in which an emotional responses was so strong as to impede deliberate reasoning. Hence fear was a dangerous "passion", but so too were other strong emotional disturbances associated with "positive emotions" like love or friendship.

potentially serious pathologies, and thus of staying healthy. At first, she may be tempted to refuse the blood-draw but, if she ponders the situation long enough, she might instead conclude that "being scared" does not justify renouncing to the test. Once the short-time distress of the blood-draw is compared with the long-term benefit of knowing valuable health information, it might appear to her that if her goal is to preserve her well-being, then undergoing the blood test is for her the right thing to do.

Of course, she could also decide otherwise. There is no a priori way of deciding which option is the right one *in general*: this evaluation always depends on several variables–including the personal value one attaches to "not being scared" and the relative importance of the information returned by the test. The final choice is always up to the autonomous judgment of the patient, and thus different persons may end up making very different and equally legitimate choices in this situation. As already stressed, Ethical Counselling should never be directive: its goal is not that of pushing the patient towards one choice instead of another, but rather that of ensuring that she acquires the ability to properly justify whatever choice she ends up making.

The crucial point is that one has the chance to deliberately choose whether or not this test is worth taking only if she can also reflect on the options at stake. Even to refuse the test because one is "too scared" is in principle a viable and legitimate solution as long as it is the outcome of a process of rational moral deliberation. By contrast, running away in fear would directly undercut one of the options, and in any case it would hardly count as one's "deliberate choice". Engaging in reason-giving is thus important because it allows agents to appraise which among the several courses of possible action best conforms to their rationally endorsed goals, thus opening the possibility for them to override their first emotional responses.

3 Reason-Giving and the Making Explicit of Personal Philosophies

Reason-giving implies a dialogic exchange between the counsellor and the counselee finalized at better articulating the moral reasons structuring an ethical dilemma. As it will become apparent in later chapters, in many cases the articulation of moral reasons consists in the identification and appraisal of a series of arguments in favour of or against a certain conduct. So, to make one example in Giovanna's case part of the ethical dilemma concerns whether or not she has to disclose certain information to her children. Accordingly, the following reason-giving process focuses on the conflicts between considerations related to the respect of patient's autonomy versus other considerations related to non-maleficence and beneficence. In this narrow sense, what we have so far defined as a "reason-giving" can be roughly equated with "argument analysis", the hallmark activity of contemporary clinical ethics and moral philosophy.

Understood as "argument analysis", reason-giving is an essential component of Ethical Counselling both for clinicians and for patients. For both methodologies, the identification and analysis of abstract arguments, as well as the definition of the key terms comparing in them, is a necessary part of what the counsellor and the counselee (patient or clinician) are expected to do during a typical session. At this point, however, it is also important to add that by "reason-giving" we also intend something more than a shared discussion about some typified and impersonal arguments. In order to clarify this crucial point, we shall focus on explaining the role of reason-giving with respect to the articulation of patient's Personal Philosophy. Understanding this broader sense in which we use the term "reason-giving" is crucial to understand our methodology for ethical counselling.

As presented in chapter "Ethical Counselling for Patients", when the counselee is a patient, the aim of the counselling is not that of clarifying an abstract ethical dilemma *per se*, but that of clarifying a specific ethical dilemma in the light of patient's Personal Philosophy. In order to solve the dilemma at hand, it is not enough for the patient neither to have a sufficient understanding of the abstract structure of similar ethical dilemmas, nor to have a clear picture of all the clinical implications of her choice. Instead, what it is also required is for the patient to consider and appraise the situation from the point of view of her Personal Philosophy, that is to say, from the point of view of that more or less defined and coherent set of beliefs, experiences and values defining her personal world view.

The crucial question is then how should a counsellor proceed to unveil patient's Personal Philosophy. Our answer is that the counsellor should again engage in reason-giving. In this case, however, reason-giving is not primarily directed at analysing abstract and impersonal arguments, but rather at *making explicit* patient's Personal Philosophy. Thus, in the *ethical assessment phase*, the counsellor probes patient's Personal Philosophy through the skilful use of *why*-questions. These questions encourage the patient to state *her moral reasons*, thus making explicit *why* she thinks that she is facing an ethical dilemma; *why* she thinks that neither of the options is ideal from a moral point of view; *why* she thinks that performing an certain action would be right or wrong in this and that circumstances, etc. *Why*-questions are important because they allow focusing the conversation on what counts as a "moral reason" in the patient's world view. By identifying such elements, the counsellor can progressively unveil patient's Personal Philosophy in relation to her ethical dilemma.

Of course, different patients may identify different kinds of "reasons". For example, in front of the same clinical situation, different persons may provide different "reasons" for why they should not undergo a certain medical procedure. For example, patients may state: "I fear the pain"; "My mother underwent the same procedures and she suffered from complications"; "I feel that it would not be 'me' anymore after this procedure"; "Since *that* experience, I distrust conventional medicine and doctors", and so on. It is at this level that patients' narratives, feelings and other aspects of their Personal Philosophies are made explicit and examined with respect to the ethical dilemma at issue. Importantly, here the counsellor's role is not that of ruling out certain reasons in favour of others, but that of guiding the patient in *making explicit* all the relevant reasons that in her view pertains to the ethical dilemma.

Making explicit patient's Personal Philosophy is always important because it may clarify why a given dilemma arose in the first place. For example, the need to undergo a blood transfusion would cause an ethical dilemma only for someone self-identifying as a Jehovah witness, but not for someone endorsing a different belief-system. In these cases, making explicit patient's Personal Philosophy allows to better understand why a given ethical dilemma arose in the first place.

Furthermore, examining patients' Personal Philosophy may directly allow overcoming their "decisional paralysis".[2] It could be that the primary cause of patients' uncertainty concerns their Personal Philosophy, rather than the moral dilemma itself. One can be in a decisional paralysis not only because one fails to determine which *means* best conform to her *ends*, but also because one can be unsure about what her *ends* are. This may occur because people seldom reflect systematically on their innermost values, goals and beliefs. It is thus possible that prior to ethical counselling one's Personal Philosophy is mostly implicit or that, once made explicit, it appears incoherent.[3] Furthermore, ethical dilemmas happen in close proximity to life-changing events that are likely to modify one's identify and life priorities. Therefore, there are cases in which patient's "decisional paralysis" may depend on their Personal Philosophy being too implicit, partially incoherent, or yet undetermined.

Whenever the counsellor suspects that patient's "moral paralysis" depends primarily on the status of her Personal Philosophy, she can further engage the counselee with more targeted questions aimed at making explicit those aspects of her Personal Philosophy that are still implicit, incoherent, or yet to be fully developed. In these cases, the counsellor's aim is that of allowing the patient to become more aware of her Personal Philosophy, and thus eventually able to figure out which of the available courses of actions (the *means*) best conform to her (now clarified) world view (the *ends*). In this respect, sometimes the simple fact *of stating* one's Personal Philosophy in front of someone else may yield unexpectedly positive results.

However, making explicit patient's Personal Philosophy may not be enough to overcome one's moral paralysis. It is still possible that someone with a clear and already explicit Personal Philosophy may nevertheless be uncertain about how to cope with an ethical dilemma. For example, it might be unclear how one's Personal Philosophy applies to the specific case at hand. So, while Giovanna may state that

---

[2]With respect to the methodology presented in chapter "Ethical Counselling for Patients", this further analysis may take place both in the ethical assessment phase and in the ethical comparative phase, where the patient is lead to elaborate her value-ranking.

[3]For example, one may states that "natural medicine" is for her "the only viable option" for healing, and yet admit that if a young child needs surgery, then one has good reasons to proceed, even if this requires the administration of antibiotics.

her primary goal is that of "protecting her children", she might still be unsure about how to specify this general principle in the present situation. In other cases, instead, it might be unclear what the options of stake are, or whether one set of reasons clearly offset the other. In other words, one might know perfectly her *ends*, and yet be unsure about how those *ends* maps on the available *means*, or about what the nature and the implications of the *means* are. More in general, human morality is very complex, and so each problem and situation may present its own distinctive features.

If the making explicit of patient's Personal Philosophy does not overcome the moral paralysis, the counsellor might engage in reason-giving as to further *articulate* patient's moral reasons. In this second sense, reason-giving is thus increasingly directed towards the making explicit of the *inferential structure* and *consequences* of patient's moral reasons, rather than their *premises*. At this stage, reason-giving may increasingly assume the form of "argument analysis", thereby focusing more on clarifying the key-terms at stake, on the inferential development of patient's moral reasons–seeing to what logical conclusions they might lead to–, or on examining the situation through the lens of exemplar cases or thought experiments.

Ideally, reason-giving culminates with moral deliberation. Proper moral deliberation occurs when the counselee forms a "considered opinion" about the ethically problematic decision at stake. This "considered opinion" consists in a rational evaluation of all the available courses of action in the light of the goals and values that the patients would deliberately choose to endorse. A convenient test to know whether the counselee has formed a sufficiently "considered" opinion is to ask her whether she would be ready to explain and defend her conclusion in front of someone asking her for the "reasons" of such choice. (To some extent, this includes as well the possibility of fully articulating why one is still uncertain about how to best resolve the decisional impasse at hand). If the answer to this question is affirmative, then the ethical counsellor has successfully reached her goal with respect to the case at hand.

To summarize, the *making explicit* of patients' Personal Philosophy is important under three respects. First, ethical dilemmas may arise as a direct consequence of having a particular Personal Philosophy. Second, the status of "moral paralysis" may sometimes depend on one's having a too implicit, incoherent or undetermined Personal Philosophy. In these cases, a moral choice turns into an ethical dilemma because the agent has no clear perception of her ends, and thus cannot determine which one of the two alternative courses of action ought to be eventually pursued. Third, making explicit patient's Personal Philosophy is required in order to exercise practical wisdom, i.e. using reason to decide which course of action best conforms to one's goals, deliberating accordingly.

4 Reason-Giving as an Open-Ended Process

Reason-giving allows clinicians and patients to better deal with ethical dilemmas. Thus, the key insight of our approach is that, if properly conducted, the making explicit of the moral reasons surrounding an ethical dilemma may in itself contributes in important ways to its understanding and possible solution. However, as a general methodology, *reason*-giving has also some limitations, the most important of which is that it does not guarantee the solution of ethical dilemmas. Three factors may concur in determining this less-than-ideal outcome:

(i) *The dilemma does not have a simple, straightforward solution.* Some dilemmas may not have a simple and straightforward solution. Sometimes engaging in *reason*-giving allows the counselee to realize that the ethical dilemma is just a terminology issue. Other times, instead, the making explicit of one's Personal Philosophy clearly indicates which of the two courses of action ought to be pursued. Yet, there are also cases in which a careful appraisal would reveal not only that the dilemma is genuine, but also that the two sets of conflicting reasons are almost equal in force, so that it is still unclear what one ought to do.

(ii) *There is not enough time/resources.* Reason-giving requires times as well as other cognitive resources to properly occur. The amount of time required by each case depends on several factors, such as the difficulty of the dilemma at hand, the skills of the counsellor, and the previous knowledge of the counselee. Hence, the time allocated for ethical counselling may sometimes be insufficient to reach a properly pondered moral deliberation.

(iii) *The patient does not want to change her Personal Philosophy.* Not everyone is open to the possibility of making explicit or revising her beliefs, values and ideas. For some patient, certain values, for example religious ones, may be non-negotiable, and thus reason-giving may not be conducted beyond a certain point.

These limitations highlight that it is better to think about reason-giving as an open-ended process, rather than as a methodology that guarantees a certain result.

Nevertheless, engaging in reason-giving may be useful even if it does not allow to solve an ethical dilemma. First, reason-giving allows both patients and clinicians to better understand the structure of the ethical dilemma at hand. Second, the very fact of engaging in reason-giving with a counsellor may aid the counselee in avoiding or mitigating the impact of some cognitive biases and logical fallacies that she would otherwise fail to notice. As we explain in the next chapter, this is for example the case of people reasoning on probabilistic issues, where different ways of framing the same choice may induce people to alter their decisions in irrational ways. Therefore, even if the session of ethical counselling ends with the counselee having still doubts about how a dilemma could be solved, engaging in reason-giving at least minimize the chances that such a doubt is a genuine one and not due to a misunderstanding of the clinical situation, the status of one's Personal Philosophy, or other factors that might lead astray our process of rational and deliberate thought.

References

Avramova YR, Inbar Y (2013) Emotion and moral judgment. WIREs Cogn Sci 4:169–178

Brosch T, Scherer KR, Grandjean D, Sander D (2013) The impact of emotion on perception, attention, memory, and decision-making. Eur J Med Sci 143:w13786

Haubner B, Dwyer S, Hauser M (2009) The role of emotion in moral psychology. Trends Cogn Sci 13(1):1–6

Phelps EA, Ling S, Carrasco M (2006) Emotion facilitates perception and potentiates the perceptual benefits of attention. Psychol Sci 17(4):292–299

Phelps EA (2004) Human emotion and memory: Interactions of the amygdale and hippocampal complex. Curr Opin Neurobiol 14(2):198–202

Prinz J (2006) The Emotional Basis of Moral Judgments. Philos Explor 9(1):29–43

Wheatley T, Haidt J (2005) Hypnotic disgust makes moral judgments more severe. Psychol Sci 16:780–784

The Centrality of Probability

Giovanni Boniolo and David Teira Serrano

Abstract This chapter deals with a topic whose importance is too often ignored with respect to Ethical Counselling: probability. Probability, indeed, is at the core of many ethical decisions encountered in the age of molecular medicine, as in the case, for example, of carrier tests or predictive and presymptomatic tests, or whenever survival rates are at issue. Thus, understanding correctly the probabilistic information is extremely important and crucial and an ethical counsellor cannot be unprovided with such knowledge.

Keywords Ethical Counselling · Probability · Diagnosis · Cognitive biases

Patients' decisions often, if not always, follow the communication of a *possible* onset or of a *possible* development of a disease. Thus, possibility, i.e. probability, is at the core of any decisional process in particular of any ethical decisional process concerning not only diagnostic or therapeutic paths but even how to plan life. It means, on the one hand, that patients should understand the probabilistic information provided to them by the physicians, and, on the other hand, that physicians should have understood the meaning of those probabilistic statements they communicate: a requirement, the latter, that seems not so satisfied considering the results of a study made by Wegwarth et al. (2012), according to which the majority of the American oncologists does not have a great awareness of the meaning (i.e. the philosophical foundations) of the statistics they use on a daily basis.

G. Boniolo (✉)
Dipartimento di Scienze Biomediche e Chirurgico Specialistiche, University of Ferrara, Ferrara, Italy
e-mail: giovanni.boniolo@unife.it

G. Boniolo
Institute for Advanced Study, Technische Universität München, Munich, Germany

D. Teira Serrano
Departamento de Lógica, Historia Filosofía de La Ciencia, Universidad Nacional de Educación a Distancia (UNED), Madrid, Spain
e-mail: dteira@fsof.uned.es

© The Author(s) 2016
G. Boniolo and V. Sanchini (eds.), *Ethical Counselling and Medical Decision-Making in the Era of Personalised Medicine*, SpringerBriefs on Ethical and Legal Issues in Biomedicine and Technology, DOI 10.1007/978-3-319-27690-8_6

Actually, physicians should know such a meaning, since it is a part of their duty. Patients should understand physician's communications, since it is matter of their life. However, what about the ethical counsellors? They should have at least a minimal knowledge and understanding of these mathematical tools, since it is impossible to help someone to disentangle a moral dilemma involving probabilistic issues without having any hint on them. An ethical counsellor has to be aware of what the clinician should know and of what the patient might ask to know.

In what follows, we address the issue of probability as it arises in two extremely pervasive cases: predictive and presymptomatic molecular tests and survival rates.

1 Probability and Tests

1.1 Some Basic Hints

As said, let us consider a predictive or a presymptomatic molecular test. This is a laboratory procedure aimed at detecting a particular molecular marker (e.g. the prostate-specific antigen, the mutated BRAC1) that allows to assess the probability that a given individual could be affected by a particular disease (e.g. prostate cancer, breast cancer). According to the *Biomarkers and Surrogate Endpoint Working Group* of the American National Institute of Health, a biomarker is "objectively measured and evaluated as an indicator of normal biological processes, pathogenic processes or pharmacologic responses to a therapeutic intervention" and it has been established "on epidemiological, therapeutic, pathophysiological or other scientific evidence".

There are two points that should be immediately emphasized. *First*, a molecular marker is not a cause, but the indicator of a physiological situation, which could be a cause: the high level of PSA is not the cause of the prostate cancer, but the indicator of a physiological situation that could be causally linked with prostate cancer. *Second*, the detection of a molecular marker via test is not per se a fully diagnostic tool, but a way of "suspecting" that there could be, or there will be, a disease. To have a complete and reliable diagnosis, usually, other evaluations have to be made. Actually, it is something telling us whether that individual is more or less prone to have a given pathological situation.

Since we have spoken about causes, it should be recalled the difference between *deterministic causality* and *probabilistic causality*. In the first case, any time we have a causal event, we necessarily have the effect event. For example, any time we have the mutation of the Huntingtin gene (the causal event), we deterministically have the Huntington's disease (the effect event). That is, the probability of having a Huntington's disease given the mutation of the Huntingtin gene is equal to 1:

$$P(\text{Huntington's disease/mutation of the Huntingtin gene}) = 1.$$

Here, the causal event is both necessary and sufficient.

In the second case, any time we have a causal event, we increase the probability of having the correlated effect event. For example, the detection of a MYH7 mutation (the causal event) increases the probability of having a hypertrophic cardiomyopathy (HCM) (the effect event). That is, the probability of having a hypertrophic cardiomyopathy given the MYH7 mutation is greater than the probability of having a hypertrophic cardiomyopathy given the non-mutated MYH7:

$$P(\text{HCM/mutated MYH7}) > P(\text{HCM/non mutated MYH7}).$$

In this case, the mutated MYH7 is necessary but not sufficient, due to the incomplete penetrance.[1] The same goes when we claim that the HCM probabilistically causes the sudden cardiac death (SCD). That is,

$$P(\text{SCD/HCM}) > P(\text{SCD/non HCM}).$$

Note that in this case, we have a neither necessary nor sufficient condition (SCD can be caused by something different from HCM, for example, by coronary artery abnormalities, or by long QT syndrome).

There is another point to clarify and it concerns *risk factor*, which, according to the World Health Organization (WHO), "is any attribute, characteristic or exposure of an individual that increases the likelihood of developing a disease or injury". Usually, a risk factor is not a cause, but it has to do with underlying causes. For example, we know that breast cancer risk factors are gender (being a woman) and age (growing older), but neither gender nor age is the cause. More clearly, to be an Ashkenazi woman is not a breast cancer risk factor because being Ashkenazi causes breast cancer, but because of the prevalence[2] of the mutated BRCA1 and BRCA2 genes (the probabilistic causes) in the Ashkenazi female population (these mutated genes are likely to be about 10 times more common in Ashkenazi Jewish women than in other female population).

1.2 The Probability of a Disease

Let us suppose that an individual goes to the clinician and the latter decides to prescribe a test to check the presence of a suspected pathological situation. The individual could ask: "If I test positive, do I have the disease?". In order to answer, the physician should know and have understood some statistical elements. In

---

[1]The *penetrance* is the percentage of individuals carrying a particular variant of a gene (allele or genotype) that expresses an associated trait (i.e. phenotype), which can be also a pathological trait. The penetrance is *complete* if all individuals who have the disease-causing mutation have clinical symptoms of the disease. It can be *incomplete* if some individuals do not express the trait, even though they carry the allele.

[2]The *prevalence* is the percentage of a population found to have a condition (a disease or a risk factor). It differs from *incidence*, which is a measure of the new cases arising in a population over a given period (month, year, etc.). Prevalence is an answer for: "How many individuals in this population have this disease now?". Incidence is an answer for: "How many individuals in this population acquire this disease per month/year?".

particular, (i) the tabulation of the population with respect to that test and that disease; (ii) Bayes theorem and what is involved in it.

First of all, any test is performed with a given *sensitivity* and with a given *specificity*. *Sensitivity* is defined as the probability to test positive given the presence of the disease. *Specificity* is defined as the probability to test negative given the absence of the disease. Let N be the number of the individuals (the population) on which the test has been made and that has developed (or not) a fully diagnosticated disease. We could summarize the results in the contingency table below.

	D	wD	
T^+	a	b	$a + b$
T^-	c	d	$c + d$
	$a + c$	$b + d$	$N = a + b + c + d$

where D = individuals with the disease; wD = individuals without the disease; T^+ = individuals test positive; T^- = individuals test negative; and a, b, c, d = number of individuals characterized by that physiological situation (D or wD) and by the test outcome (T^+ or T^-). Thus, we have that:

$$\text{Sensitivity} = P(T^+/D) = a/(a+c)$$
$$\text{Specificity} = P(T^-/wD) = d/(b+d)$$

Of course the test could not be perfect, and therefore, there could be a fraction of the *false-positive* results (the test is positive, but there is not the disease) and a fraction of the *false-negative* results (the test is negative, but there is the disease). That is,

$$\text{False positive fraction} = P(T^+/wD) = b/(b+d) = 1 - \text{Specificity}$$
$$\text{False negative fraction} = P(T^-/D) = c/(a+c) = 1 - \text{Sensitivity}$$

Thus, a clinician prescribing a test should know its sensibility and its specificity (and therefore, the false-positive and the false-negative fraction correlated with it).

Already at this point, the clinician could answer a question that could be raised by the supposed individual seats in front of him/her: "What is the probability that I really have the disease if I am tested positive?", or "What is the probability that I do not have the disease if I am tested negative?". That is, the clinician could calculate the probability of having the disease given the positive outcome of the test (*positive predictive value*), or the probability of not having the disease given the negative outcome of the test (*negative predictive value*).

$$\text{Positive Predictive Value} = P(D/T^+) = a/(a+b)$$
$$\text{Negative Predictive Value} = P(wD/T^-) = d/(c+d)$$

Actually, these probabilities are meaningful only with respect to the prevalence of the supposed disease in the population of interest: another datum that the clinician should know. The higher the prevalence of the disease in the considered population is, the higher the positive predictive value (and the number of positive cases detected, i.e. the *yield*) is. Note that while sensitivity and specificity are characteristics of the test, positive and negative predictive values depend on the prevalence of the disease in the population of interest.

There is, however, a better way of putting all together both the information we have (on the test and on the population of interest) and the correlation between positive/negative predictive value and prevalence. This way is offered by the powerful and elegant Bayes theorem.

It is not relevant here to know how Bayes theorem has been formulated and proved. Instead, it is important that it allows the clinician to connect the data concerning the sensitivity and the specificity of the test with the prevalence of the disease in an explicit manner. Moreover, it is a formal tool that permits to move from the pathological level concerning the knowledge of the effects given the causes (synthesized in $P(T^+/D)$) to the clinical level concerning the knowledge of the causes given the effects (synthesized in $P(D/T^+)$).

Abstractedly, it claims that

$$P(H/E) = \frac{P(E/H) \cdot P(H)}{P(E)}$$

where

- $P(H/E)$, called the *posterior probability*, is the conditional probability of the hypothesis H given the evidence E, and it gives the degree of belief in H, having taken E into account;
- $P(E/H)$ is the conditional probability of E given H;
- $P(H)$, called the *prior probability*, is the probability of H alone, and it gives the initial degree of belief in H;
- $P(E)$ is the probability of E alone.

In our case, the probability of having the disease D (the hypothesis) if the test T (the evidence) is positive—that is, $P(D/T^+)$—and considered the prevalence of the disease in the population of interest (i.e. the positive predictive value) becomes

$$P(D/T^+) = \frac{P(T^+/D)P(D)}{P(T^+)} = \frac{\text{sensitivity} \cdot \text{prevalence}}{a + b/N}$$

where, as said, sensitivity = $P(T^+/D)$ and prevalence = $P(D)$.[3]

Now, it is immediate to understand why $P(D/T^+)$ is a probability a posteriori (it is the probability of the disease after the test) and $P(D)$ is a probability a priori (it is

---

[3]Of course if $P(D)$ = prevalence, then $P(wD) = 1 - P(D) = 1 -$ prevalence.

the probability of the disease before the test). It is to note that since sensitivity and prevalence are at the numerator of the fraction, the higher they (independently) are, the higher the positive predictive value is.

Easily, we could also know

(1) the negative predictive value, that is, to be tested negative without having the disease

$$P(wD/T^-) = \frac{P(T^-/wD)P(wD)}{P(T^-)} = \frac{\text{sensitivity} \cdot (1 - \text{prevalence})}{c + d/N}$$

(2) the probability of a false positive, that is, to be tested positive without having the disease:

$$P(T^+/wD) = \frac{P(wD/T^+)P(T^+)}{P(wD)} = \frac{(1 - \text{sensitivity}) \cdot (a + b/N)}{1 - \text{prevalence}}$$

Let us illustrate what above considering the case of a test detecting a marker for colorectal cancer, i.e. the M2-PK.[4] Let us suppose that it is a very reliable test with a sensitivity of 85 % (i.e. 0.85) and a specificity of 95 % (i.e. 0.95). We know also that the cumulative risk of colorectal cancer in persons aged under 75 is 3.9 % (i.e. 0.039) in Europe. At this point, we can construct our contingency table, considering a population of 100,000 individuals and calculate both the positive and the negative predictive values.

	D = individuals with colorectal cancer (confirmed with colonoscopy)	wD = individuals without colorectal cancer (confirmed with colonoscopy)	
T^+ = M2-PK tests is positive	2797 (a = true positive)	4835 (b = false positive)	7632
T^- = M2-PK test is negative	493 (c = false negative)	91,875 (d = true negative)	92,368
	3290	96,710	N = 100,000

$$\text{Positive Predictive value} = \frac{\text{sensitivity} \cdot \text{prevalence}}{\frac{\text{True Positive} + \text{False Positive}}{N}} = \frac{0.85 \cdot 0.039}{\frac{7632}{100000}} = \frac{0.033}{0.076} = 0.43 = 43\ \%$$

$$\text{Negative Predictive value} = \frac{\text{sensitivity} \cdot (1 - \text{prevalence})}{\frac{\text{False Negative} + \text{True Negative}}{N}} = \frac{0.95 \cdot (1 - 0.039)}{\frac{92368}{100000}} = \frac{0.913}{0.924} = 0.99 = 99\ \%$$

---

[4] It is the dimeric form of the pyruvate kinase isoenzyme type M2.

Now, we could ask: "Is the test clinically relevant?". We know that it depends on the prevalence, since it depends on the difference between the knowledge we have before the test to be affected by the disease as members of a population and the knowledge we have after the test to be affected by the disease. In the case above, the probability before the test to have a colorectal cancer was 3.9 %. After the test, we know that if the test is positive, we have a probability of 43 % of developing it, while if the test is negative, we have the probability of 99 % of non-developing it. That is, we have a real increase of knowledge and the test is worth being done.

2 Probability and Survivorship

A *survival analysis* deals with any positive or negative end point event (discharge from the hospital, graduation, marriage, first metastasis, death, etc.). It is used in cancer studies to plot the development in time of the disease in a group of individuals by taking into account tumour types (sites, histology), period of diagnosis, gender, stages, ethnicity, etc. Here, the end points of major interest are patients' death and cancer recurrence. In particular, *survival rates* (and their synthesis in time: the *survivor curves*) show the percentage of individuals with a certain type and stage of cancer who has survived the disease for a certain period of time after they have been diagnosed.

Usually, it is indicated a *5-year absolute survival rate*, which refers to individuals who are alive at least 5 years after they have been diagnosed. Of course 5-year is a conventional period which could be useful in aggressive cases (lung cancer) but which could be less useful in non-aggressive cases (prostate cancer). It should be distinguished from the *5-year relative survival rate*, which describes the percentage of patients that are alive 5 years after the diagnosis divided by the percentage of the general population (with the same characteristics of gender, age, ethnicity, etc.) that are alive after 5 years. It is to note that a 5-year relative survival rate may also be equal to or even greater than 100 %, if patients have the same or even higher survival rates than the general population (it could happen that the cancer patients are cured or that they are more socio-economically privileged than the general population).

What makes the survival analysis complex is that not all the members of the group are observed for the same amount of time. First, there is problem of the patients diagnosed near the end of the study period. And it is plausible that they are alive at the last contact. Of course, in this situation, we do not know whether their survival is just as long as, or longer than, the others under observation. Second, there is the problem that it is not said that we know the outcome status of all of the patients who were in the group at the beginning. Some may be lost to follow-up for many reasons: they might have moved, changed name or physician, etc. Moreover, some could have died for reasons which have nothing to do with their oncological pathology (a car crash, a home accident, a cardiovascular disease, etc.).

Let us call *uncensored cases* the patients who are observed until they reach the end point of interest (e.g. recurrence or death), while let us call *censored cases* all the others (patients who survive beyond the end of the follow-up or who are lost to follow-up). The survival analysis leading to the construction of the survival curve has to be taken into account the censored cases properly.

The two more used method to pursue the analysis are the *life table (or actuarial) method* and the *Kaplan–Meier method*. In the first, the total period over which a group is observed is divided into fixed intervals, usually months or years. For each interval, the proportion surviving to the end of the interval is calculated on the basis of the number known to have experienced the end point event (e.g. death) during the interval and the number estimated to have been at risk at the start of the interval. For each succeeding interval, a cumulative survival rate is given by the probability of surviving the most recent interval multiplied by the probabilities of surviving all of the prior intervals. Thus, if the percentage of the patients surviving the first interval is 90 % and is the same for the second and third intervals, then the cumulative survival percentage is 72.9 %.[5] The Kaplan–Meier method calculates the proportion surviving to each point that a death occurs, rather than at fixed intervals. The cumulative survival rate is calculated rather similarly to the actuarial method.

The starting time for determining survival of patients depends on the purpose of the study. It could be the date of diagnosis, first visit to physician, hospital admission, treatment initiation, etc. It is not a specific day, but it is the time each patient enters the study. At the starting time, however, all the patients have to be alive and the survival rate is 100 %. With the flow of time, the survival rate decreases until 0 %, if the period is long enough and all the patients are dead.

In the pictures below, there are three different ways of depicting data concerning survivor analysis (it is an example concerning liver and biliary tract cancers, realized by the SEER—Surveillance, Epidemiology, and End Results—programme of the US National Cancer Institute) (Figs. 1 and 2; Table 1).

Note that in the last table, there is the *median survival*. It tells us how long 50 % of the individuals have died. It is undefined whether 50 % of the patients are still alive at the end of the study.

There is a last comment that has to be made. The *survival rates do not predict* what happens to a specific *individual*, since any patient is unique and both treatments and responses vary greatly. They could be just an indication of what has occurred to patients with that cancer. They should not become something disheartening or threatening. Moreover, the *survival rates do not evaluate the effectiveness of different forms of treatment*, which can only be determined by a properly conducted clinical trial.

---

[5]Since 90 % is 0.9, $0.9 \cdot 0.9 \cdot 0.9 = 0.729$, that is, 72.9 %.

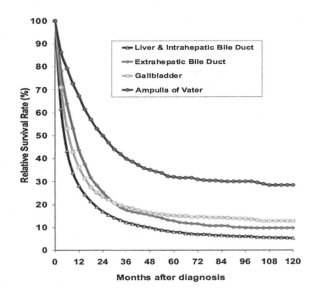

Fig. 1 Relative survival rates (%) by primary site (liver and intrahepatic bile duct, extrahepatic bile duct, gallbladder, ampulla of vater) and months after diagnosis, ages 20+, 12, SEER areas, 1988–2001

3 What About Patients?

One major finding of psychologists about our understanding of probability is that we tend to misinterpret it systematically. Gigerenzer et al. (2007) presented the following scenario to 160 gynaecologists: "Given a breast cancer screening mammogram with a sensitivity of 90 %, a false-positive rate of 9 % and a disease prevalence in the relevant population of 1 %, if a woman tests positive, what are her chances of having breast cancer?" The answers were surprisingly: 60 % of the gynaecologists believed that 8 or 9 out of 10 women who tested positive would have cancer. If we apply Bayes theorem, as explained above, we will see that this is mistaken: the probability of having breast cancer given a positive mammogram is the sensitivity of the test (90 %) times the disease prevalence (1 %) divided by the probability of obtaining a positive in the test (true or false). This is the *positive predictive value*, $P(D/T^+)$, that we calculated above and it is about 10 %. Grasping it is difficult because $P(T^+)$ is not easy to infer from the other data.

Gigerenzer and his team argue that physicians (and patients alike) are better at solving the problem when the statistical data are presented in the form of *natural frequencies*. If we think of 1000 women, a 1 % prevalence means that 10 are expected to have breast cancer. Nine of these 10 women will test positive in the mammogram (this is the 90 % sensitivity). About 89 of the remaining 990 women without cancer will still test positive (the 9 % false-positive rate). When the problem is presented with these frequencies instead of percentages, most of the gynaecologists easily grasped that the probability of obtaining a positive in the test is the sum of the 9 true positives and the 89 false positives. Of these 98 women, only 9

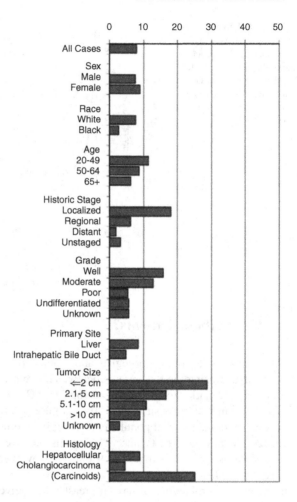

Fig. 2 Cancer of the liver and intrahepatic bile duct: 5-year relative survival rates by sex, race, age, historic stage, grade, primary site, tumour size, and histology, ages 20+, 12 SEER areas, 1988–2001

will actually have breast cancer. So the probability of having breast cancer after a positive mammogram is about 1 out of 10 (10 %).

As a consequence of such a mistake, patients who do not need treatment may end up receiving it. There are different approaches for avoiding such errors, depending on how we appraise our cognitive abilities. Some may recommend improving the statistical literacy of physicians and patients alike. But it often happens that improvements are limited, depending on the previous levels of education of the patient or of the context and manner in which the information is communicated. This is why Gigerenzer and his collaborators have been advocating for simple decision rules (*heuristics*) which would allow physicians and patients alike to handle uncertainty with a success rate similar to a complex statistical algorithm.

Table 1 Cancer of the liver and intrahepatic bile duct (excluding carcinoids): median survival time and 1-, 2-, 3-, 5-, 8- and 10-year relative survival rates by sex, race, age, historic stage, grade, primary site, and tumour size, ages 20+, 12 SEER areas, 1988–2001

Characteristics	Cases	Percentage	Median survival (months)	Relative survival rates (%)					
				1 year	2 year	3 year	5 year	8 year	10 year
All cases	13,409	100.0	4.6	28.2	17.0	12.2	8.0	6.1	5.3
Sex									
Male	9224	68.8	4.3	27.2	16.3	11.7	7.5	5.9	4.7
Female	4185	31.2	5.3	30.5	18.7	13.3	8.9	6.5	6.3
Race									
White	8619	64.3	4.3	26.8	16.2	11.6	7.7	5.9	5.3
Black	1497	11.2	3.5	22.1	11.4	7.6	2.7	1.5	1.5
Other	3293	24.6	6.0	–	–	–	–	–	–
Age									
20–49	2030	15.1	5.5	32.3	20.7	15.6	11.5	9.0	8.3
50–64	4235	31.6	5.0	30.8	19.0	14.3	8.7	6.5	5.3
65+	7144	53.3	4.1	25.4	14.7	9.9	6.3	4.6	4.0
Historic stage									
Localized	4021	30.0	9.8	46.9	33.0	25.4	18.1	14.4	12.5
Regional	3487	26.0	4.8	27.5	14.1	9.8	6.1	4.2	3.3
Distant	3299	24.6	2.8	12.3	5.4	3.2	1.8	1.1	1.1
Unstaged	2602	19.4	3.7	20.5	11.3	6.9	3.2	2.2	2.0
Grade									
Well	1717	12.8	10.3	47.5	32.0	24.3	15.8	13.1	10.7
Moderate	1860	13.9	7.0	38.0	24.9	19.2	12.9	9.0	7.5
Poor	1904	14.2	3.4	20.7	11.2	7.3	5.4	3.0	2.7
Undifferentiated	273	2.0	2.7	18.2	9.9	6.6	5.7	5.7	3.4
Unknown	7655	57.1	4.0	23.7	13.5	9.3	5.7	4.4	4.0
Primary site									
Liver	11,598	86.5	4.4	28.2	17.5	12.9	8.5	6.5	5.6
Intrahepatic bile duct	1811	13.5	5.6	27.9	14.5	8.5	4.8	3.4	3.2
Tumour size									
<=2 cm	489	3.6	18.8	59.0	47.0	39.4	28.8	23.8	17.3
2.1–5 cm	2077	15.5	10.2	47.5	33.3	24.7	16.7	13.0	10.9
5.1–10 cm	2386	17.8	6.4	36.2	21.2	15.5	10.8	7.9	6.4
>10 cm	1236	9.2	5.4	30.1	16.7	11.8	8.9	7.3	6.8
Unknown	7221	53.9	3.2	17.5	9.1	5.9	3.0	2.0	1.8

– Not calculated

Fast-and-frugal trees are an instance of these medical decision-making heuristics, even if (alas!) they have not yet applied in the field of medical genetics and oncology. The trees articulate three different cognitive tasks: (i) *searching* for information about patient prognosis; (ii) *stopping* this search as soon as the relevant information is gathered; and (iii) *classifying* it according to its relevance so that the physician can make a decision. There are a limited number of branches in the tree, constraining the number of predictors that can be looked up according to their statistical reliability, so that the subsequent decisions are made simpler. For example, the following tree provides a heuristic for deciding whether a patient should be assigned to the coronary care unit or to a regular nursing bed (Marewski and Gigerenzer 2012). The first predictor is a given anomaly in an electrocardiogram. The second one is about the patient's major complaint. Third one is a composite of five other predictors. We can assess how many correct or incorrect decisions were made with the tree as compared to unaided medical judgment, on the one hand, and a statistical model, on the other. Trees may not be as predictively reliable as a formal model and are perhaps more cumbersome than an expert clinical eye. But they are faster than the former and less biased than the latter, providing a reasonable compromise in error rates.

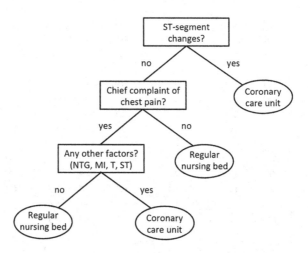

But there are even more complex forms of probabilistic biases in medical practice. If we consider again survival analysis, we may wonder whether it may help us assessing the efficacy of screening programs in preventing cancer deaths. It is true that screen-detected cancers have better 5-year survival rates than cancers detected because of symptoms, but, against the belief of many physicians, that does not imply that screening saves lives. This is the so-called *lead-time* bias. We saw above how the starting time for determining survival depends on the purpose of the study. Should it be the point at which the patient is screened or the point at which the cancer starts? Imagine a patient in whom the cancer starts at 50; the symptoms will become visible at 67, and she will die at 70. Her 5-year survival rate is 0: she dies three years after the diagnosis. But imagine now that she is screened at 60 and the cancer is detected then, even if it will inexorably kill her at 70. Here, her 5-year survival rate is 100 %. But that does not imply that screening saved her life. The lead-time bias makes us think otherwise: we need to choose the correct starting time for survival analysis (Wegwarth et al. 2012).

Lead-time bias

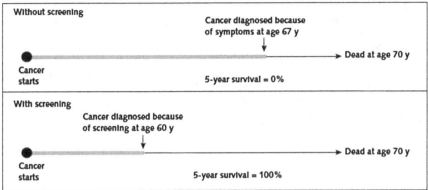

Screening may detect cancers that may have remained asymptomatic during the life of the patient. She may have even died of other causes. In such cases, screening generates overdiagnosis, a true positive for which the clinical consequences are not straightforward: the costs of treatment may not yield a significant benefit for the patient. Communicating these risks in a clear manner to the patient is an equally challenging task for which we do not have clear standards yet.

In any case, medical decisions under uncertainty are not independent of the values of the patient (his aversion to risk, for instance). The ethical counsellor should help the patient not just in understanding the true risks that he/she is facing, but also in articulating his/her own preferences regarding the consequences: e.g. the trade-off between the benefits of early detection and the losses derived from years of unnecessary treatment if overdiagnosed. *Fast-and-frugal trees* do not only make decisions simpler and more effective for physicians, but also make the process more transparent and accessible for patients. These or any other user-friendly tools are called for if we want a more evidence-based health care in which statistics improve rather than hinder our decisions.

References

Dupras C, Ravitsky V (2013) Disclosing genetic information to family members: the role of empirical ethics. In: eLS. John Wiley & Sons Ltd, Chichester. http://www.els.net

Gigerenzer G (2002) Calculated risks: how to know when numbers deceive you. Simon & Schuster, New York

Gigerenzer G (2013) Five year survival rates can mislead. BMJ 346:f548

Gigerenzer G (2014) Breast cancer screening pamphlets mislead women. BMJ 348:g2636

Gigerenzer G et al (2007) Helping doctors and patients make sense of health statistics. Psychol Sci Public Interest 8:53–96

Lautenbach DM et al (2013) Communicating genetic risk information for common disorders in the era of genomic medicine. Annu Rev Genomics Hum Genet 14:491–513

Marewski JM, Gigerenzer G (2012) Heuristic decision making in medicine. Dialogues Clin Neurosci 14:77–89

Motulsky H (2010) Intuitive biostatistics. A nonmathematical guide to statistical thinking. Oxford University Press, New York

Wassertheil-Smoller S, Smoller J (2004) Biostatistics and epidemiology: a primer for health and biomedical professionals. Springer, Heidelberg

Wegwarth OI et al (2012) Do physicians understand cancer screening statistics? A national survey of primary care physicians in the United States. Ann Intern Med 156:340–349

Wiseman M (2010) Communicating genetic risk information within families: a review. Fam Cancer 9:691–703

Part II
Ethical Issues

Genetic Testing and Reproductive Choices

Paolo Maugeri

Abstract This chapter aims to offer an introduction for the management of the ethically controversial issues arising from the relationships between reproductive choices and genetic testing. In particular, three main ethical issues recurrent within this debate are presented and analysed. First of all, it is shown that reproductive choices can be *different number choices*, meaning that the reproductive decision will change the number of individuals eventually born, or *same number choices*, meaning that the reproductive decision will not change the number of individuals eventually born. Secondly, considerations pertaining to the quality of life and welfare of the future children are introduced as having a pivotal role in the assessment of controversial reproductive issues. Finally, the principle of procreative beneficence is presented.

Keywords Ethical Counselling · Genetic testing · Reproduction · Procreative beneficence

Decisions as to whether (not) to have children are amongst those choices that mostly impact on the lives of individuals facing them. Becoming parents, indeed, involves a relevant emotional and psychological investment that maybe only few other life experiences demand. Ever since medical knowledge has extended its diagnostic powers to reproduction—from the analysis of pedigrees up to the finest scale of Preimplantation Genetic Diagnosis—the decisional landscape disclosing before prospective parents has magnified to unprecedented levels, and with it their responsibilities. To cope with this, genetic counselling is routinely offered to individuals undergoing genetic testing as a way to inform patients about the inherited disorder they may be affected, the risk of transmitting it to potential offspring, and as a support for informed decision-making. As important as genetic counselling may be, however, it alone does not exhaust the full repertoire of issues enmeshed in reproductive choices. As illustrated in the first part of this book, in fact, medical decision-making is only part of a broader scenario that involves also

P. Maugeri (✉)
Department of Experimental Oncology, European Institute of Oncology (IEO), Milan, Italy
e-mail: paolo.maugeri@ieo.eu

© The Author(s) 2016
G. Boniolo and V. Sanchini (eds.), *Ethical Counselling and Medical Decision-Making in the Era of Personalised Medicine*, SpringerBriefs on Ethical and Legal Issues in Biomedicine and Technology, DOI 10.1007/978-3-319-27690-8_7

considerations recruiting patients' moral perspectives and dilemmas. The latter is the primary focus of Ethical Counselling.

Accordingly, in this chapter we shall provide a tool for orienting around the most relevant ethical issues and arguments arising from difficult reproductive choices prospective parents are, sometimes, required to make. In particular, we will proceed as follows: first, we shall provide a brief overview of the genetic procedures available for prospective parents; second, we shall outline the concept of "pro-creative liberty" as the background against which the ethical analysis should begin; and finally, we shall elaborate on a conceptual road map—characterized by three ethically relevant landmarks—that should help the counsellor to analyse various reproductive choices with regard to their ethical underpinnings. As a way of illustration, we will also show how the YY syndrome case (Case 4, The Plan) is taken into account in our conceptual road map.

1   Genetic Tests: A Brief Overview

In the context of reproduction, genetic tests are used as an aid for prospective parents in order to identify, or exclude the presence of, DNA mutations that are associated with genetic diseases.

In order to support parental choices, genetic tests can be performed on the prospective parents or in the embryo/foetus. In the first case, the test is done by analysing a small sample of patients' blood or saliva so as to assess, before pregnancy, the risk of passing on a genetic disorder to their offspring. In the case of the future born, instead, genetic testing can be divided into three broad categories: postnatal testing, prenatal testing, and preimplantation genetic diagnosis.

Postnatal testing works along the same lines indicated above, and it is used to establish whether the newborn is affected by genetic disorders, like for instance cystic fibrosis, and eventually device a therapeutic path when available.

Prenatal testing means all those techniques that are used to diagnose a vast array of (genetic and not genetic) diseases in the foetus. A non-exhaustive list is given as: Down syndrome, spina bifida, Tay-Sachs syndrome, thalassaemia, sickle cell anaemia, fragile X syndrome, muscular dystrophy, etc. Prenatal diagnosis can be performed by non-invasive and invasive methods. Amongst the former, we have: *ultrasonography*, an examination of the woman womb through ultrasounds in order to assess, among other things, the gestational age, the growth and sex of the foetus, and the presence of a twin pregnancy; *maternal blood screening*, which consists in the analysis (between the 15th and 18th week of gestation) of dosage levels of alpha-fetoprotein, beta human chorionic gonadotropin, and free estriol, in order to evaluate the risk for the foetus to develop a number of pathologies; maternal blood screening can a be used also as a way to detect foetal DNA in order to identify chromosomal aberrations, such as Down syndrome. Amongst the invasive tests, instead, we have: *amniocentesis*, which consists in the collection (around the 15th week of gestation) of a sample of amniotic fluid in order to detect a number of conditions (such as Down syndrome, cystic fibrosis, muscular dystrophy, genetic deafness, haemophilia); *villocentesis*, which consists in the collection (around the 10th week of gestation) of a sample of chorionic villi and overlaps with amniocentesis with regard to the number of identifiable conditions.

Preimplantation genetic diagnosis (henceforth PGD) is a procedure complementary with prenatal tests and is offered to prospective mothers in the context of assisted reproductive technologies (henceforth ART), and more specifically of in vitro fertilization (IVF). PGD is done on the embryo before it is implanted in the uterus, and it overlaps with amniocentesis and villocentesis in terms of the identifiable conditions. Differently from the latter, however, PGD can be used—within the limits established in different national legislations—as a means to select "healthy" embryos for implantation.

2 The Nature and Scope of Procreative Liberty

Before starting with the ethical analysis, it is worth introducing the concept of *procreative liberty* as it will prove helpful in setting the stage for the scrutiny of ethically controversial reproductive choices. It is important to notice at the outset that the aim here is to present the scope of the concept, and not its normative force. Namely, we are not arguing that this liberty should always trump on other moral considerations. Rather, we aim to illustrate how far this concept extends so to better grasp how other relevant moral concerns (such as, for instance, parental responsibility, duty of care, newborn's quality of life) may play out with, or eventually infringe, it. In the bioethical literature, the concept of "procreative liberty" is taken to include a number of capacities for prospective parents to self-determine along a number of dimensions that include the following (Buchanan et al. 2000, 209–212):

1. *whether to procreate with a willing partner.* This capacity will include activities and choices that lead or support reproduction (i.e. access to ART and genetic information), and activities that are meant to prevent it (i.e. access to contraception).
2. *when to procreate.* In reproduction, timing is crucial, as fertility diminishes with age. Thus, this capacity will include access to techniques that allow managing the timing of pregnancy (i.e. cryopreservation of the gametes), or postponing it to an age that would have been unlikely to reproduce even a few decades ago (i.e. through access to ART).
3. *how many children to have.* This capacity is sometimes infringed for reasons of demographic control in countries such as China, but it is not of particular concern for our purposes.
4. *what kind of children to have.* This is the most controversial part or procreative liberty, especially when it implies possible harm to the child. For instance, genetically deaf parents may ask to access IVF and PGD to ensure that their kid will inherit their disability, as it happened in one case much discussed in the

bioethical literature (see, for instance, Savulescu 2002). As controversially, some prospective parents may ask, in the context of PGD, to select embryos for non-medical conditions (i.e. high IQ, sex, and other conditions).

5. *whether to have biologically related children.* This capacity refers to the uncoerced access to ART, included techniques of heterologous fertilization (i.e. gametes or embryo donation, and maternal surrogacy).

The moral underpinning of reproductive liberty is the strong interest human beings have in *self-determination,* namely the idea that it is of high value to be the authors of those choices that profoundly affect our future and our identity, and to do that without external coercive forces. In this respect, reproductive choices are certainly among the most significant and fundamental ones. Despite its importance, self-determination alone does not cover all the morally relevant considerations that have a bearing on reproductive choices. More specifically, the procreative liberty of prospective parents may well be counterbalanced (at least on a purely moral level) by beneficence-based considerations regarding the newborn's welfare or, even infringed, when they can cause harm to the child. Ethical Counselling should be very sensitive to these dimensions when clarifying and investigating patients' Personal Philosophy. In particular, when approaching a patient, the ethical coun-sellor should first clarify in which of the capacities outlined above, if any, the particular case she is analysing is more problematic. For instance, if the case she is analysing is one of a couple accessing ART for heterologous fertilization, the relevant capacity will be the fifth in the list outlined above, while she will take into account the second if the case at hand is one of a woman accessing ART at a relatively advanced age.

With regard to Julie's case, and similar ones, the capacities that require attention are the second (when to procreate) and the fourth (what kind of children to have). As for the second capacity, it must be noticed that Julie and Philip are in their forties when they access ART. In a case like this, the ethical counsellor could explore together with the couple, the motivations behind their choice and the moral values underpinning it, and see whether there are any that could eventually lead to decision paralysis as they proceed. The fourth capacity is the most problematic one in the case at hand. In particular, here the couple is choosing not to be aware of a piece of information that could prove important in raising up the child and giving him the best educational support for the full development of his potential. This issue will be developed in some more depth in Sect. 3.3, for the moment it is important to remark, however, that Julie's choice does not seem to involve any morally relevant harm to the child, at least not one that trumps her entitlement to procreative liberty. As we will see in the following sections, a number of other considerations may partake to reproductive decisions for which ethical counselling support is required.

3 Mapping the Ethics of Reproductive Choices

Having explored the concept of procreative liberty, it is now time to dig a little bit deeper in the ethical issues potentially arising in reproductive choices. On this issue, scholars in bioethics have exercised in a considerable debate that it is impossible to reconstruct in its full details here. Therefore, rather than embarking in this daunting task, we will provide the ethical counsellor with a conceptual road map that should serve as a reference guide to subsume the particular case she is confronting with into more general ethically relevant categories. In doing this, we shall also show, as a way of illustration, how to place Julie's case into the frame here offered, leaving to the next two chapters the analysis of the ethical issues peculiarly emerging from the case at hand.

3.1 The First Landmark: Same Number Versus Different Number Choices

The first landmark of our conceptual road map for the ethical assessment reproductive choices concerns the recognition of the number of individuals the latter will impact on. In this respect, reproductive choices can be: (i) *different number choices* —the reproductive decision will change the number of individuals eventually born (i.e. abortion); (ii) *same number choices*—the reproductive decision will not change the number of individuals eventually born (i.e. PGD cases). These distinctions, although couched in rather obscure philosophical terminology, are nonetheless important to distinguish among the vast variety of reproductive choices and ethical implications thereof.

To make this distinction apparent, let us think of two couples pursuing two different paths of ART. The first couple conceives the kid "naturally" (i.e. through coitus) and then the woman undergoes amniocentesis, the test results in a diagnosis of Down syndrome. As a consequence of the diagnostic outcome, the couple considers abortion for a number of reasons (for instance, because they do not feel ready to cope with the needs of a kid affected by this condition). The second couple, instead, undergoes PGD on the embryos obtained through an IVF cycle. One of the embryos detected by PGD displays a chromosomal aberration (e.g. Down syndrome). As a consequence, the couple decides to implant the "healthy" embryo, discarding the others.

At a first reading, one may be lead to think that the two cases present similar if not identical ethical implications, them being different only with regard to the specific Personal Philosophies subscribed by the actors at play (i.e. the prospective parents). On the contrary, the two cases bring about different sets of moral consideration that is worth enucleating separately.

In particular, in different number choices, the morally relevant option against which the reproductive choice has to be made is between bringing into existence an individual with, a more or less severe, disease or disability, or no individuals at all. In same number choices, instead, the option is between bringing into existence a future individual (that is, an individual with Down syndrome), with her potential personal identity, array of life plans and quality of life, or a different individual (that is, an otherwise healthy individual) with her own peculiar set of opportunities and quality of life. In the former case, one of the morally relevant issues—all other things being equal—has to do with the question whether a life with disability has to be preferred, in aggregate terms, to no life at all. In the latter, the question changes dramatically to whether bringing about a healthy life is—all other things being equal—better than bringing about one beset with disability. The problem here is not only terminological, but it brings with it substantial moral implications. In different number choices, the implications can be subsumed under the general heading of the morality of abortion, thus recruiting relevant considerations as to when abortive choices are morally acceptable. Among them, disability diagnoses are generally regarded as the most stringent reasons for pregnancy termination, right after serious health threats for the woman and before personal reasons in postponing pregnancy for pursuing career development goals. In same number choices, instead, ethical analyses need to incorporate specific considerations on the identity affecting implications of selection, like: is there a duty to implant the best embryo? If not, are prospective parents who decide to implant embryos with genetic disorders harming them?

Some of these issues will be clarified in the following paragraphs. With these distinctions in mind, it is now time to see how to frame Julie's case in this respect. Since she is not undergoing PGD, hers is not a case of same number choice and thus it qualifies properly as a different number one. However, she never considers abortion a viable nor a desirable option, thus she promptly solves any different number *empasse*. A different scenario would have been opened up if Julie had undergone IVF and PGD and, on the diagnosis of one embryo affected by the YY syndrome, she had chosen to implant it anyway, may discarding others. In this reconstructed scenario, Julie's case would properly be described as a same number choice, thus carrying with it all the above-mentioned implications. Are there reasons to problematize the ethics of this newly described choice? To answer to this question, it is necessary to further crawl into our conceptual road map in the search of other morally relevant concerns.

3.2   The Second Landmark: Considerations on the Quality of Life of the Child

The discussion developed so far let emerge a first distinction to be drawn for the ethical assessment of controversial reproductive choices. The different lines we put forth, however, left much of the issues underspecified. In order to enrich the analysis is now worth introducing considerations concerning the quality of life of future child. The latter marks a morally relevant difference in potential reproductive choices and, in this context, they represent the second landmark in our conceptual road map.

In bioethics, considerations pertaining to the quality of life and welfare of the future children have pivotal role in the assessment of controversial reproductive issues. Among the various distinctions proposed by the scholars active in the field, a general one deserving attention is that between cases of interventions and choices compatible with a worthwhile life and those that are not such. Although many people may find this locution disquieting, it encapsulates an array of intuitions worth exploring. In a standard text in the field the distinction is presented as follows:

> A life not worth living is not just worse than most people's lives or a life with substantial burdens; it is a life that, from the perspective of the person whose life it is, is so burdensome and/or without compensating benefits as to make death preferable (Buchanan et al. 2000, 224).

This passage conveys the idea that certain existences are besets with conditions, for instance Tay-Sachs, so serious such to be incompatible with any prospect for self-determination. In other terms, even from the perspective of the individuals affected by those conditions, their lives would be so painful as to be preferable not to be born at all. If this line of reasoning works, then there would also be strong moral reasons for prospective parents to avoid bringing into existence individuals with such a low level of welfare.

Many people would resist this intuition and maintain that, on the contrary, any life is always worth living. No matter how much pain it may involve; it nonetheless would include some, even intangible, element of joy that would redeem sufferance. For instance, a person affected by an invalidating disability may very well keep herself alive on other grounds, for instance for benefiting others (Wilkinson 2010, 70). Accordingly, at least a couple of objections are generally placed against the idea that the ethics of reproductive choices is better explained in terms of the welfare of the future born. One strategy points to the idea that ARTs, at least those done with the intent of selecting who will eventually be born, run contrary to a truly ethical conception of parenting. The latter would, instead, better be conceived in terms of a gift (Sandel 2007). Refusing children with disability, so the argument goes, displays a bad attitude towards life, which is something we should always cherish and value rather than attempting to dominate and control. As a consequence, all lives should always be welcomed with joy and responsibility, and no choice based on purely *welfarist* grounds is ever ethically acceptable. Another strategy stems from the perspective of disabled people themselves. This objection is termed *expressivist*, in that it argues that diagnostic techniques such as amniocentesis or PGD are ultimately displaying an ugly attitude towards disabilities. For instance, a couple deciding to terminate pregnancy after finding out that the foetus is affected by Down syndrome, is sending, more or less explicitly, the message that the world would be a better place if only individuals with Down syndrome will not be around. A couple of counter-objections can, however, be placed against these readings. If on the one hand is true that the ethical dimension of parenting cannot be reducible to the idea of liberty, and it certainly includes care and commitment, it is not clear, on the other, why the idea of giftedness should be considered the right enriching metaphor for the ethics of parenting. Human beings display robust preferences and, among these, avoiding diseases for themselves and their relatives is one of the most robust. Accordingly, it is hard to see how using preimplantation and prenatal diagnostic tools would display a bad attitude in parenting more than other more traditional standards of care. Moreover, while the expressivist argument actually enriches our understanding of disability, it is not apparent the causal link between the choice of prospective parents to abort a diseased foetus or to select healthy embryos for implantation with the stigmatization of disabled people. As Glover (2006, 35) puts it:

> We need to send a clear signal that we do not have ugly attitudes to disability. It is important to show that what we care about is our children's flourishing […] To think that a particular disability make someone's life less good is not one of the ugly attitudes. It does not mean that the person who has it is of any less value, or is less deserving of respect, than anyone else.

The second landmark helps us clarifying some issues touched at the end of the previous paragraph. With regard to the YY case, we can tell that it is certainly an instance of *life worth living*. The YY syndrome, indeed, is a mild condition bringing about, with a 50 % probability, only minor impairments in learning and language acquisition, sometimes coupled with ADHD (that is, Attention Deficit Hyperactivity Disorder) symptoms. The latter, moreover, can be easily managed through appropriate educational interventions. Similar considerations can be put forward in the reconstructed scenario that sees Julie undergoing IVF and PGD. Upon her choice to implant an embryo displaying an YY karyotype, it is hard to advance a harm-to-the-child argument. Indeed, the only alternative for that embryo, if not implanted, would have been non-existence and not an existence without the condition. Slightly different considerations could instead put forth if the condition under scrutiny would not qualify as straightforward case of a life worth living. In those circumstances, the ethical counsellor may explore the Personal Philosophy of the patient along the lines here developed.

3.3   The Third Landmark: The Principle of Procreative Beneficence and the Child's Right to an Open Future

The last concept we present as a landmark in our conceptual road map is the so-called principle of procreative beneficence. This principle states that in reproductive choices prospective parent should act in order to maximize the welfare of their children. This means that, when genetic tests are available, couples have a *prima facie* moral obligation to have the best child, of the possible children they can have, with the best prospects of life (Savulescu 2001, 2002). This principle can be given different interpretations. A first, and strongest, interpretation requires that the moral obligation arising from the principle is actually a binding one. That is, parents *ought* to maximize life prospects of the child. A second version would, instead, interpret the moral obligation stemming from the principle in a light manner, conceiving maximization as a laudable act and not as a strict moral injunction. Both versions are, however, problematic in that they would demand, in a more or less stringent manner, to extend the purview of required parental actions for the benefit of the child to the whole spectrum of genetic interventions. Moreover, this would be true both for interventions aimed at disease avoidance and for the positive selection of non-medical traits (such as, for instance, height, eye colour).

The implications of such a principle, especially in its strong interpretation, are therefore far-reaching and amenable to a number of remarks. On a positive ground, the principle seems to operate upon morally laudable intuitions. Indeed, choices aimed at welfare maximization would, in general, align the interests of the future born, of the parents and, perhaps, of society. It seems plausible to say that, all other things being equal, a child would prefer to be healthy rather than be impaired due to a genetic condition; at the same time, both the family and the society would in general prefer to deal with a fully operative and healthy member. Besides running contrary to the idea of parenting as a gift outlined above, this line of thought lends itself to other objections, of these we will cite two.

A first objection is based on the concept of procreative liberty. In this regard, it seems a violation of procreative liberty to require too high an investment of prospective parents on their offspring, especially when this implies sacrificing all energies from their personal development or other obligations they already have in place. This objection seems to work with the strongest interpretation of the principle and only insofar as it is taken to implausibly imply complete and exclusive devotion to the child. With specific reference to genetic interventions, moreover, it is not clear why choosing the best possible child would necessarily impose unbearable burdens to the prospective parents so to undermine the exercise of their procreative liberty.

A second objection to the principle, instead, is based on another entitlement future born would have, namely the so-called child's right to an open future. The latter has been presented by the legal scholar Feinberg (1980) as a comment to a famous sentence of the US Supreme Court in the case *Wisconsin v. Yoder* (1972). The problem was whether Amish communities requiring their children to leave school before 16 (as required by the law) as to enhance their integration in their community of origin were violating a right of the child to acquire those capacities necessary for choosing and developing sufficiently vast array of life options normally available in society. In the case of genetic interventions, especially those aimed at the enhancement of non-medical traits, the child's right to an open future may be undermined when prospective parents operate with the aim of imposing specific characteristics on their children as to make them inclined to pursue specific life plans. Whether this line of reasoning actually counts as a decisive objection to the principle of procreative beneficence is open to debate, for instance, it is not clear why choosing for healthy traits should ever be seen as a limitation in this respect.

As some commentators remark (Buchanan et al. 2000) health is an *all-purpose good*, namely one that is useful, if not sometimes fundamental, for the development of any life plan. Accordingly, choosing for healthy traits would rarely be inconsistent with the right to an open future.

The concepts here presented are also relevant for the analysis of the YY syndrome case. Indeed, we know that Julie has decided not to test the twins to make sure of who of them was affected by the condition. A reading of this choice may suggest that in so doing she is partly limiting the open future of one of her kids. Indeed, knowing who is bearing the condition may help Julie and her family to device appropriate educational plans so to disclose to the child a broader array of life options, perhaps maximizing his capacities to pursue the life plan he may see suit. On the other hand, Julie may nonetheless offer a vast array of educational means to her children—for instance, through close and constant scrutiny of their capacities and developmental progresses—though lacking this bit of information. As we have seen, indeed, the YY syndrome is not a major condition, in that it does not severely impair affected individuals who, given appropriate support, grow normally and are able to overcome issues in learning and language acquisition that may eventually arise. More specifically for the case, it is open to further scrutiny— to be explored in the concrete and unique situation of Ethical Counselling—whether Julie's choices (and alike) are substantially limiting the child's right to an open future and, in the case they do, whether this would count as morally relevant limitations. Indeed, the issue opens up another significant question that may counterbalance the considerations developed in this paragraph. Namely, is Julie exercising a putative right not to know? On what grounds can we justify her waving of this information? These are issues we will develop further in the following chapters.

4 Conclusion

In this chapter, we offered to the ethical counsellor a tool for orienting herself in the intricate moral issues arising from reproductive choices. Our goal was to provide a conceptual road map for framing concrete cases into more general categories that can help systematizing ethical analysis. In doing this, after exploring the nature and scope of procreative liberty, we have identified three landmarks that would serve as reference points for enucleating and clarifying the different moral intuitions and arguments that we may find at play in reproductive choices. As a way of illustration, we also placed Julie's case into this general frame and made some remarks on how this played out in this respect. In the following chapters, we will further put under scrutiny the case at hand through the analysis of a putative right not to know and the ethical issues arising from incidental findings.

References

Buchanan A, Brock DW, Daniels N, Wikler D (2000) From chance to choice. Genetics and justice. Cambridge University Press, Cambridge

Feinberg J (1980) The child's right to an open future. In: Feinberg J (ed) Freedom and fulfilment: philosophical essays. Princeton University Press, Princeton, pp 76–97

Glover J (2006) Choosing children. Genes, disability and design. Clarendon Press, Oxford

Sandel M (2007) The case against perfection. Harvard University Press, Cambridge

Savulescu J (2001) Procreative beneficence: why we should select the best children. Bioethics 15:413–426

Savulescu J (2002) Deaf lesbians "designer disability" and the future of medicine. BMJ 325:771

Wilkinson S (2010) Choosing Tomorrow's Children. The ethics of selective reproduction. Oxford University Press, Oxford

The 'Right-not-to-Know'

Luca Chiapperino

Abstract This chapter deals with the ethical controversies related to the so-called right-not-to-know in biomedicine. By problematizing the substantive conflicts at the basis of patient's decision to waive some health-related information, the chapter provides a normative map for instructing the practice of Ethical Counselling in the face of these claims. In particular, it is argued that both self-regarding and other-regarding considerations in the exercise of the right-not-to-know may ground or dismiss its ethical and legal legitimacy and may prove fundamental aspects of ethical counselling processes.

Keywords Ethical Counselling · Genetic testing · Right-not-to-know · Duty-to-inform

Do patients have a right to waive information and knowledge about their health conditions? This issue, usually presented under the label of one's 'right-not-to-know', has been the focus of much philosophical and bioethical scrutiny in the last three decades. Yet, the recent expansion in the availability of genetic technologies for diagnostic and prognostic purposes has provided the case of information waivers with renewed bioethical piquancy.[1] As we will see in the next sections of this chapter, both international guidelines and bioethics scholars often assume that the right-not-to-know (one's genetic predisposition to illness) should be respected. Usually, these arguments are grounded on considerations of

---

[1]This is not meant to affirm that genetic information differs, in an ethically relevant manner, from other types of medical information. Rather, this chapter will treat genetic knowledge very much like any other kind of knowledge about our health, thus rejecting claims of 'genetic exceptionalism'. For a discussion of the reasons in favour and against genetic exceptionalism, see Green and Botkin (2003), Rothstein (2007).

L. Chiapperino (✉)
Institute of Social Sciences, University of Lausanne, Lausanne, Switzerland
e-mail: luca.chiapperino@unil.ch

© The Author(s) 2016
G. Boniolo and V. Sanchini (eds.), *Ethical Counselling and Medical Decision-Making in the Era of Personalised Medicine*, SpringerBriefs on Ethical and Legal Issues in Biomedicine and Technology, DOI 10.1007/978-3-319-27690-8_8

welfare for the patient, namely the fact that some kind of information may potentially harm psychologically the patient without producing any particular benefit, and therefore, it is in the best interest of the patient to waive such knowledge. Nonetheless, most contributors to this debate agree that a number of competing ethical considerations should come into play when taking into account claims in favour of the right-not-to-know.

The case of Julie (Case 4, The Plan) provides, in this respect, a rich example on which to draw most of these considerations that are of utmost importance also for the practice of Ethical Counselling here introduced. First, Julie is faced with the difficult decision of refusing to know some information that may be of substantive relevance to her child(ren), thus providing us with an occasion to discuss the limits and restrictions we might ascribe to claims in favour of the right-not-to-know. Second, her case is particularly fertile because it adds another dimension to the controversies internal to the debate on the right-not-to-know. In Julie's case, the information regarding the YY condition of one of her twins is in fact not only relevant to her and her beloved, for the sake of planning an adequate education and support to the development of their child(ren). Rather, the developing child is also a *prospective stakeholder* of the right-not-to-know his genetic condition, which her mother is asked to exercise. The particular case we will analyse in this chapter constitutes therefore a good example of how the often-assumed right-not-to-know may assume less cogency when third parties are involved, and in particular when children are the subjects potentially affected by the information to be waived.

To show this, we will first turn to the bioethical literature dissecting the normative foundations of claims in favour of and against the 'right-not-to-know'. This analysis will set the stage for an application of the methodology for Ethical Counselling proposed to the particular case of Julie, unravelling the ethical intricacies of her decision not to test her children.

1   The Nature and Scope of One's 'Right-not-to-Know'

Dealing with the nature and scope of what is generally called the 'right-not-to-know' entails analysing what are the conditions for respecting claims in favour of not knowing some health-related information, and—in recent writing—one's genetic make-up. As we will see in this section, such claims depend on the trade-off between different ethical concerns, values in biomedicine, and on a number of factual conditions about the information (potentially) waived by the patient.

Many bioethical scholars have rightly pointed out that there is a legitimate (often presumed) interest, on the side of individuals, in not knowing some (genetic) information. Even though informed consent (one's right-to-know about the medical procedures she is subjected) is widely regarded as the main pillar of contemporary medical ethics, it is generally agreed that its protection of patient autonomy carries with it at least two corollaries. First, patients should be allowed to refuse any treatment since, by denying them this possibility, the very same idea of consent would be meaningless. Second, patients have also a correspondent 'right', or an entitlement to refuse knowing some information about themselves and their own health. As briefly mentioned above, this latter interest of patients is often justified by *welfarist* considerations[2] regarding the impact that unrequested, or scarcely manageable, information may carry for patient's coping with his/her medical condition. According to this line of reasoning, daily clinical practice presents us with scenarios in which knowledge may simply be too burdensome for the patient (e.g. knowledge of genetic predispositions to diseases) and has a detrimental impact over her life as well as her well-being. In such situations, it may therefore be preferable to allow these people the possibility of refusing to receive such information, thus protecting their quality of life. Alternatively, another consideration in favour of the second corollary to the golden rule of informed consent comes from the status of *autonomous* agent ascribed to patients. Inasmuch as waiving information is the direct result of one's autonomous decision-making, any patient ought to be granted a right *not* to be informed.

Nevertheless, it is important to stress that this legitimate interest of patients should not be taken for granted. Neither it has absolute value, nor it has to be treated as an uncontroversial ethical matter. Rather, the right-not-to-know is conditional to other morally worthy considerations, which undermine the *prima facie* substantive justifications for waiving information grounded on patient welfare and autonomy.

1.1 Autonomy and Its Conflict with the 'Right-not-to-Know'

The most fundamental objection that can be moved to claims in favour of the right-not-to-know resorts to the value of knowledge for autonomy and moral agency. This chapter is, of course, no place to delve into the complexities of the philosophical tradition on the concept of *autonomy*.[3] It will suffice to say for our purposes that, if we conceive autonomy—the defining feature of moral agents—as the expression of rationality and freedom of the will, we are also bound to acknowledge the relevance of knowledge in guiding moral reasoning; namely, we ought to recognize that it is likely to make better decisions on the basis of better information and knowledge. This implies that refusing some information is against

---

[2]By "welfarist considerations" we mean here to refer, very broadly, to the justificatory nature that affecting the well-being of an individual may have for guiding human agency. For a philosophical introduction to the concept of well-being and its normative import on welfarist moral theories, see Roger Crisp's entry "Well-being" in the Stanford Encyclopedia of Philosophy.

[3]Very much like in the case of 'welfarism' above, the reader who may want to plunge into the philosophical debate on the concept of autonomy may find useful to consult John Christman's "Autonomy in Moral and Political Philosophy" in the Stanford Encyclopedia of Philosophy.

our rational will (i.e. irrational) and therefore *ipso facto* a non-autonomous decision, which undermines our status and capacities as full-blown moral agents.

Taken to its extreme implications, the reasoning presented here affirms that, granting a right-not-to-know to a patient, means allowing such agent to deny (or to jeopardize at least) its moral status as an autonomous being. And by the same token, supposing that we would go as far as saying that only moral agents (i.e. autonomous ones) are indeed right-holders, the criticism above would mean that granting a right-not-to-know to a patient is indeed to vindicate the contradictory claim that a right lies with a non-right-holder.

This autonomy-based rejection of the right-not-to-know is not meant to deny that alternative justifications could be found for accepting patient decision to waive a particular bit of information. Recall that, indeed, autonomy-based claims in favour of such right are only one of the two justifications we introduced for justifying such entitlement. For instance, all of the welfare-based assessments of the right-not-to-know are left untouched by the deconstruction of autonomy as its moral basis. Even by knowing that we are letting someone acting against her status as an autonomous agent by waiving a piece of information, we might in fact still be prone to recognize that, if such information is detrimental to her well-being, it is morally worthy to let her do so. Furthermore, we might also point out that the autonomy-based rejection of the right-not-to-know presented above may be too extreme of an argument. One might reply in fact that autonomy is not a monolithic attribute of a person and that we ought to resist rushing to the conclusion that an individual who waives *some* information is irrational and non-autonomous. We, as human beings, do plenty of irrational and inconsistent deeds, and yet this seems no sufficient ground to advocate against our (limited) freedom to indulge in such behaviours.

Otherwise stated, this critique of the right-not-to-know is best understood as providing compelling philosophical reasons only to deconstruct the apparently incontestable nature of such claims, rather than their absolute acceptability. First, this criticism has the merit of distinguishing between opposite value orientations that can be identified in the actions of information waivers. The other side of the coin of autonomously renouncing to get informed about one's health conditions is a conflict that arises with autonomy itself and with the value of knowledge for exercising it. Second, the autonomy-based critique of the right-not-to-know also warns us as to the potential limitations of right talks in discussing ethically sensitive matters. Although the language of 'rights' resorts to strong intuitions about what we morally owe to one another, the critique introduced above alerts us as to the insidious implications of this language. The uncritical acceptance of a right-not-to-know overlooks that exercising such a right means in part undermining the foundations of any right we bear, namely our status as rational and free agents.

1.2 The Externalities of the 'Right-not-to-Know': A Revival of Paternalistic Medicine

A second criticism of the right-not-to-know focuses on its potential impact on the doctor–patient relationship. Building upon its autonomy-based rejection, some authors have argued that an additional reason not to admit patient's waiving of information lies in the importance of protecting professional 'duty-to-inform' them about avoidable risks and relevant health information.

From this perspective, the by-product of recognizing anything such as the right-not-to-know would be eroding those developments in biomedical ethics aimed at moving the doctor–patient relationship from a paternalistic and compliance-oriented model towards a full appreciation of patient choice and autonomy as the ultimate guidance in medical decision-making. Besides therefore having an obligation towards ourselves (as autonomous agents) to be fully informed when we make decisions, this second critique points to the other-regarding aspect of exercising the right-not-to-know. Namely, to the obligation we have towards present and future patients not to foster a culture of paternalism in health care, whereby responsibility for important decisions is left in the hands of professional authority, and patients are devoid of—or expected to refrain from—choosing what is in their best interest.

1.3 Third-Party Interests and the 'Right-not-to-Know'

Another limitation to the right-not-to-know comes from the consideration of the potential interests in the information waived, residing on third parties such as patient's relatives. This issue is particularly salient in the case of germline genetic testing, where the discovery of a particular mutation (e.g. TP53 gene mutation that increases predisposition for a number of cancers, as illustrated in Case 1, The Plan) is relevant knowledge not only for the individual tested, but also for his/her family members. Test results are in fact likely to highlight genetic predispositions to disease shared with patient relatives, thus presenting the issue of whether to inform family members about those serious risks, or susceptibilities bearing upon their health. Within this context, it is generally agreed that the professional may play a pivotal role. Even though healthcare personnel is bound to a duty of confidentiality, it seems hardly defensible to argue that they should conceal to patient relatives an information that can greatly interest and benefit them. In particular in those cases where a preventative measure, or a treatment, is available for the tested condition, it seems reasonable to conclude that a professional duty to beneficence trumps the claim in favour of the right-not-to-know. Things stand differently, however, in cases where any beneficial measure is absent for third parties, and thus, the presumption of their strong interest in knowing the information waived is no ground for breaking confidentiality and privacy.

Another set of considerations regarding third-party interests deals with cases of proxy decision-making (such as Julie's case), where the individual concerned by the information to be waived is unable to decide for him/herself. Frequently, these scenarios involve parents who decide to have their children tested for a genetic condition (e.g. chromosomal aberrations, or tests for the estimation of genetic risk for cancer). Most of the authors in the bioethical literature agree upon the fact that, being incompetent agents, children should be subjected to these tests only for the sake of their best interest and that the responsibility for managing this information lies in the hands of parents. Nevertheless, a strong disagreement exists about how to define and enforce a meaningful conception of 'best interest' in this situation.

This controversy has in fact ignited a passionate debate, which can be summarized in the opposition between two different stances. One strand of thinkers has adopted a *prohibitive stance* justified by the concern that, if no readily medical benefit is available for the condition at issue, this information has no particular utility and might even psychologically harm both the parents and the (developing) child. In addition, proponents of this view have also argued that waiving this kind of knowledge about the child has the additional merit of protecting his/her future right to decide—as an autonomous adult—about whether to know this information. Otherwise stated, the 'prohibitive stance' has the merit of protecting the child's decision (as a prospective stakeholder) to relinquish knowledge about herself, namely his/her *future* right-not-to-know.

Different scholars have instead emphasized an *open stance* on this matter, by arguing that parental exercises of the right-not-to-know aimed at protecting the future autonomy of the child are indeed problematic, if not contradictory. The first reason to conclude this could be found in the obligations and duties of parents *qua* parents and proxy decision-makers. According to this line of reasoning, a couple refusing information that is relevant for the child's health could be said to be acting against their parental responsibility, which compels them to identify the best possible means to promote the well-being and autonomous development of the child. Achieving such a goal may be in fact very difficult on the basis of missing, or incomplete information about the child's health and biological predispositions. By the same token, relinquishing an information about one's child is also violating their obligation as proxy decision-makers to come to the best possible choice about treatment and health measures to be adopted for the child. Also in this case one might question their capacity to come to the best decision in the interests of the child, by starting from partial knowledge of his/her health. Secondly, different scholars have argued that waiving information about the child's health can hardly be reconciled with the idea of protecting her future autonomy and right-not-to-know, because this knowledge may actually be fundamental (or at least significant) for the development of the child's autonomous life plan. Getting to know (for instance) some genetic predispositions, they argue, is an important bit of individual self-knowledge, partaking in the development of one's conception of a good life. From adequate education and support to the growth of the child, to decision-making concerning reproductive choices, health-related and genetic information has a life-changing effect over the way in which parents may raise their child, and over the identity and life plans she will construct for her adulthood. Otherwise stated, if knowledge regarding risk of developing some diseases bears upon decisions that make our lives plain (e.g. reproductive choices of young women who are found to be positive to risk-enhancing BRCA1/2 mutations), then why should we withhold this information to children? This knowledge, one might argue, is as important to them (as developing agents) as it is to any other agent.

Also in the case of balancing the right-not-to-know with the moral interests of third parties, a number of moral considerations seem to counter the *prima facie* acceptability of granting the exercise of such right. Striking the balance between the ethical stances outlined above is, however, not an easy task for the analyst, or (as we will see) for the counsellor. This choice depends in fact very much on weighing what medicine has to offer for a particular condition against the value choice, pertinent to individual Personal Philosophy, of prioritizing present or future (like in the case of children) autonomy of third parties affected by punctual exercises of the right-not-to-know.

2 The Legal Recognition of the 'Right-not-to-Know'

Before putting to work the previous analysis of the normative foundations of the right-not-to-know in the context of Julie's case, it may be useful to clarify also what kind of legal recognition this entitlement has come to acquire in the last few decades. As we learn from Andorno (2004), a number of national and international documents recognize a *presumption* of right-not-to-know as one of their fundamental principles, even though a number of excusing conditions to forbid its exercise are also instated by these documents.

A useful example of such documents is the European *Convention on Human Rights and Biomedicine*[4] (Art. 10.2), which states 'Everyone is entitled to know any information collected about his or her health. However, the wishes of individuals not to be so informed shall be observed'. Otherwise stated, according to the Convention, individual right-to-know goes hand in hand with her right-not-to-know, which is justified by personal reasons that should not be *in general* subject to the scrutiny of healthcare authorities. As clearly affirmed by the Explanatory Report[5] of the Convention, "Patients may have their own reasons for not wishing to know about certain aspects of their health. A wish of this kind must be observed".

Yet, at the same time, the Convention identifies also the need to draw a limit for the exercise of the right-not-to-know. As stated in Art. 10.3 "In exceptional cases, restrictions may be placed by law on the exercise of the rights contained in Art. 10.2 in the interests of the patient". But what are the 'exceptional cases' under which waiving information is not deemed acceptable under EU legislation? The

---

[4]Council of Europe (1997a).
[5]Council of Europe (1997b).

Explanatory Report constitutes again the place where a better qualification of the provisos of the Convention can be found. In particular, the Report identifies two criteria (drawn from Art. 26 of the Convention) defining the excusing conditions for not respecting individual right-not-to-know: (i) when the restrictions are in the interests of patients' health, such as in cases of therapeutic necessity or in cases where the knowledge at stake may be the only way to enable them to take potentially effective (preventive or therapeutic) measures; (ii) when 'certain facts concerning the health of a person who has expressed a wish not to be told about them may be of special interest to a third party'. In such cases, the interests of the third party should warrant his/her right-to-know taking precedence over the patient's right to privacy and the resulting right-not-to-know.

Similar recognition of the right-not-to-know can be found also in the UNESCO *Declaration on the Human Genome and Human Rights*[6]—whose Art. 5c suggests "the right of each individual to decide whether or not to be informed of the results of genetic examination and the resulting consequences should be respected"—and the World Health Organization *Guidelines on Ethical Issues in Medical Genetics and the Provision of Genetic Service.*[7] In particular, this latter document provides a two-tiered justification of claims in favour of the right-not-to-know. On the one hand, the Guidelines evoke a welfare-based justification of such claims, thus pointing to the potential psychological and social harms that unrequested disclosure of genetic information may cause to the individual. On the other, the right-not-to-know finds recognition here also on the basis of considerations of voluntariness (as a necessary condition for any medical practice), autonomy and privacy. Yet, also a number of excusing conditions for not granting this right are stated by these Guidelines. First, the document suggests that action should be taken, on the side of genetic counsellors, in the case of third-party interests in the information to be waived, especially when a serious burden or harm can be avoided to others. Second, the Guidelines point also to the limitations of the right-not-to-know that apply to contexts of reproductive choices (where a duty can be instated to share with partners relevant genetic information, such as the status of carrier of a serious genetic condition) and to cases of prenatal, newborn and children testing for treatable conditions.

3   Is Julie and Philip's Choice not to Test the Twins Ethically Acceptable?

How could the above-mentioned considerations regarding the ethical and legal limitations of the right-not-to-know inform an ethical evaluation of the case of Julie? What should be the focus of ethical counsellors in supporting Julie's decision to waive the information as to the YY syndrome affecting one of her twins?

---

[6]UNESCO (1997).

[7]World Health Organization (1998).

As mentioned above, dealing with these questions means first and foremost to assess the value of knowledge regarding the YY syndrome for the health of Julie's developing child. The effects of having an extra Y chromosome can be very different, but the vast majority of boys (it is a sex-specific aberration) affected by this syndrome lead normal lives. As reported by a fact-sheet of Genetic Alliance UK,[8] these children "go to ordinary schools, have successful careers, marry, have children and live until old age". In many cases in fact people with an extra Y chromosome never know of their chromosomal abnormality, as no particular symptom, nor treatment, for this condition exists. They usually grow taller than the average (average height for them is 1.88 cm) and do not display any particular visible phenotype. It is reported in the literature that some people with YY syndrome may display learning difficulties, as well as behavioural problems that generally fade away with age. The motor development of YY children is usually normal (they learn to crawl and walk at the usual time), but data report a delay in the linguistic and speech development that can be usually compensated with adequate support and therapy. Children with YY syndrome have often levels of IQ lower than the average (even though they can also have exceptional levels of intelligence) and can be affected by autism spectrum disorders (ASD) as well as attention-deficit hyperactivity disorder (ADHD). Yet, they often do well at completing upper-secondary education, especially when adequately supported, and do not have problems in finding employment, or even starting a new family. Finally, people with YY syndrome are not more likely to pass this condition to their children than the average population.

The status of the information concerning Julie's child provides a *prima facie* justification for the legitimacy of her choice not to know which of the children is affected by YY syndrome. No treatment is available for such condition and it is possible that, for the first few years of his life, Julie's child will do as well as his brother (or any other infant). His condition may well require later on the adoption of some special measures for his successful upbringing, but Julie and her husband have all the information they need to take up such measures. Even though they cannot tell which of the twins has a chromosomal aberration, they are prepared to take up all the measures required to counter the appearance (in each of the two) of any of the difficulties related to the YY syndrome. The kind of knowledge Julie is waiving can therefore be regarded as a typical case of information for which the legal presumption of right-not-to-know introduced above applies. However, what kind of ethical considerations should the counsellor bring to the attention of Julie and her husband?

Drawing from our analysis of the philosophical foundations of the right-not-to-know, a number of considerations are in order for the practice of Ethical Counselling applied to this case. *First*, the ethical counsellor may focus on

---

[8]The fact-sheet on the XYY syndrome by Genetic Alliance UK can be found at the following address: http://www.geneticalliance.org.uk/docs/translations/english/25-xyyt.pdf (last accessed: July, 1 2015).

whether this information can actually be detrimental to Julie and her husband's well-being. Should they deem this kind of knowledge as particularly problematic or stressful, or perceive it as a potential threat to their social and psychological well-being, it may be worth (for the reasons presented above) for the counsellor to take a step back and avoid interfering with their decision.

Second, the counsellor may also point out the importance of this knowledge for future decisions concerning the child, and for the *autonomous* agency of his parents. The counsellor may recall Julie and her husband that, insofar as the child will be non-autonomous and unable to make any decision by himself, they are the ones in charge of making the best decisions concerning his life and development as a person. By doing so, the counsellor may also stress the importance of having a complete picture of the child's health predispositions and conditions, not only for the sake of being prepared to react to any possible developmental delay he may face, but also for the sake of being more attentive to cues and mildly visible symptoms requiring professional support. Otherwise stated, it is important for the ethical counsellor facing such a case to highlight that *general* knowledge about the consequences of YY syndrome (like the ones that Julie and her husband have) may not suffice for having the kind of regards and precautions that *one* of their children might need in his upbringing. By ignoring that one of the two is affected by a chromosomal aberration, the focus of these parents may be unnecessarily directed to—and potentially overwhelmed by—the search of worrisome cues in both of their children. Such a scenario, Julie and her husband should know, may easily be avoided by knowing test results for each child.

Third, another set of considerations for the counsellor may deal with the child's *prospective* interest of knowing he is affected by YY syndrome. As showed above, two main stances could be clarified to the parents for the future decision of whether to inform the child of his chromosomal abnormality. On the one hand, what we called the 'prohibitive stance' would point to the lack of clear medical benefit for undergoing the test as a reason not to test and inform the child. Already a cursory review of national and international recommendations for presymptomatic genetic testing in minors reveals, for instance, that policies mostly favour the limitation of testing for minors, unless genuine medical benefits are available for the tested condition. The same kind of reasoning applies to the YY syndrome, for which it seems meaningful to pander to parents' request of not testing the child. In addition, the 'prohibitive stance' may also point to the opportunity of preserving the *future autonomy* of the child in deciding not to know the results of the test. On the other hand, the counsellor may also emphasize that an 'open stance' on the issue may equally be worthy and find its justification in the idea that is reasonable for parents not to waive this information and to prepare themselves and the child (as he grows up) to cope with this bit of knowledge about him. This information may actually be useful for the child himself to construct his own life plan and goals, or to better cope with—and react to—the potential difficulties arising from his genetic condition. In a nutshell, this knowledge can be as important to him as it is to any other agent to develop his full potential as a person.

It falls within the realms of ethical counselling practice, as conceived by the present guide, to support couples in finding the right trade-off between all of these different ethical considerations. The general answer to the question 'Is Julie and Philip's choice not to test the twins ethically acceptable?' can only be therefore, from the perspective defended throughout this work, an unsurprising 'It depends'. Many moral interests are attached to choices such as the one of Julie and her husband, which deserve careful scrutiny and consideration on a case-by-case approach. Recall that the Personal Philosophy of the agents involved is, in the end, the ultimate guidance for disentangling the ethical intricacies of their exercise of the right-not-to-know. Yet, a more fundamental consideration seems to bridge the analysis of this case from the different perspectives deployed in this book (see chapters "Genetic Testing and Reproductive Choices" and "Incidental Findings") and represents the most likely point to be addressed by counsellors facing similar cases. At the basis of the decision as to how to settle all the competing arguments outlined above lies the idea of *parenting* that Julie and her husband bear in mind for the upbringing of their children. Some parents may privilege openness and sincerity at the basis of their understanding of their role, which may lead them develop a plan about the timing and person(s) best suited to convey test results to their children. Some others may have instead a different attitude and prefer to have the children investigate these issues by themselves, as soon as they reach a mature age. Parental attitudes ought, in other words, play a significant role in determining whether it is appropriate to disclose the information to the child and consequently also orient parental decision as to how to exercise their right-not-to-know. The task of counsellors and practitioners, in the face of these scenarios, is to understand the potential difficulties arising from prenatal testing, or testing of minors, and to support parents in the identification of their optimal solution to the ethical issue they are confronted with. The synergy between ethical and genetic counselling may be pivotal in this process, by rendering this choice and the information available to parents (or eventually the child) clearer through the reliance on structured forms of communication (see chapter "Reasons and Emotions"). Whether different outcomes of this process are ethically defensible depends, however, on the unique way of balancing all the distinct concerns analysed in this chapter in a specific and unrepeatable decision that ties up some parents and their children.

References

Andorno R (2004) The right not to know: an autonomy based approach. J Med Ethics 30:435–439

Christman J (2015) Autonomy in moral and political philosophy. In: Zalta EN (ed) The stanford encyclopedia of philosophy. Available at: http://plato.stanford.edu/archives/spr2015/entries/autonomy-moral. Last accessed 1 July 2015

Council of Europe (1997a) Convention on human rights and biomedicine. Available at: http://conventions.coe.int/Treaty/en/Treaties/Html/164.htm. Last accessed 1 July 2015

Council of Europe (1997b) Explanatory report of the convention on human rights and biomedicine. Available at: http://conventions.coe.int/Treaty/EN/Reports/Html/164.htm. Last accessed 1 July 2015

Crisp R (2015) Well-being. In: Zalta EN (ed) The stanford encyclopedia of philosophy (Summer 2015 edition). Available at: http://plato.stanford.edu/archives/sum2015/entries/well-being. Last accessed 1 July 2015

Green MJ, Botkin JR (2003) Genetic exceptionalism. In medicine: clarifying the differences between genetic and nongenetic tests. Ann Intern Med 138:571–575

Ost DE (1984) The 'right' not to know. J Med Philos 9:301–312

Robertson S, Savulescu J (2001) Is there a case in favour of predictive genetic testing in young children? Bioethics 15:26–49

Rothstein MA (2007) Genetic exceptionalism and legislative pragmatism. J Law Med Ethics 35:59–65

UNESCO (1997) Universal declaration on the human genome and human rights. Available at: http://portal.unesco.org/en/ev.php-URL_ID=13177&URL_DO=DO_TOPIC&URL_SECTION=201.html. Last accessed 1 July 2015

World Health Organization (1998) Guidelines on ethical issues in medical genetics and the provision of genetic service. Available at: http://www.who.int/genomics/publications/en/ethicalguidelines1998.pdf. Last accessed 1 July 2015

Incidental Findings

Maria Damjanovicova

Abstract Various medical tests are routinely performed in medical practice to establish or confirm a diagnosis and prescribe the right treatment. In some cases, the results of a medical test can reveal a previously undiagnosed condition, which is not related to the current medical condition and the original purpose of the test. Such results, called *incidental findings*, have sparked a significant debate on whether all of them should be reported back to the patients. In this chapter, we present the debate on the implications and management of incidental findings in the clinic, to offer a heuristic for reaching an ethically desirable outcome on a case-by-case basis.

Keywords Genomic medicine · Incidental findings · Patient-centred · Ethical Counselling · Clinical decision-making

Various medical tests are routinely performed in medical practice to establish or confirm a diagnosis and prescribe the right treatment. In some cases, the results of a medical test can reveal a previously undiagnosed condition, which is not related to the current medical condition and the original purpose of the test. Such results are referred to as *incidental findings* and have sparked a significant debate on whether all of these findings or some, and which ones, should be reported back to the patients. Some of the ethical issues raised in this debate are touched upon in previous chapters (see chapters "Genetic Testing and Reproductive Choices" and "The 'Right-not-to-Know'"). However, the incidental findings have been given a considerable space in contemporary debate on the management of personal data in biomedicine and deserve a separate section in this guide. In what follows, we will discuss incidental findings and the debate about the implications and management of incidental findings in the clinic, to offer a heuristic for reaching an ethically desirable outcome on a case-by-case basis.

M. Damjanovicova (✉)
Dipartimento di Scienze della Salute, University of Milano, Milan, Italy
e-mail: maria.damjanovicova@ieo.eu

M. Damjanovicova
Department of Experimental Oncology, European Institute of Oncology (IEO), Milan, Italy

G. Boniolo and V. Sanchini (eds.), *Ethical Counselling and Medical Decision-Making in the Era of Personalised Medicine*, SpringerBriefs on Ethical and Legal Issues in Biomedicine and Technology, DOI 10.1007/978-3-319-27690-8_9

1 Incidental Findings

In the case of twin pregnancy (Case 4 in "The Plan"), Julie, who is 41, underwent a routine procedure of amniocentesis to check if the twins she is carrying are developing as healthy babies. The outcome of this test was unusual because the twins showed different karyotypes: one of them had the normal karyotype (46, XY), while the other displayed an extra male chromosome (47, XYY). Amniocentesis serves to detect chromosomal aberrations, but the most commonly occurring syndromes that the test is aimed to detect are trisomies, like Down syndrome, Edward syndrome and Patau syndrome, all of which give rise to significant clinical implications. In this respect, the result of Julie's test should be considered as an incidental finding, since the test was performed to detect a certain set of (most common) clinical conditions and it showed an unexpected anomaly (and a rare one) in one of the twins. This is, therefore, a good example of incidental findings in prenatal genetic testing. Incidental findings are not however confined to cases similar to Julie's. Within genomic medicine, incidental findings can also occur in preconception screening,[1] population screening for disease risk,[2] the molecular characterisation of rare diseases,[3] the individualisation of treatment (particularly in cancer) and pharmacogenomics (see Biesecker 2012; Green et al. 2013).[4]

Incidental findings became an issue within genomic medicine in the transition from targeted genetic testing to genome-scale screening testing. Targeted genetic testing consisted of probes, which targeted particular sequences in the genome known to be linked to diseases for which the test was performed. For example, such is the case with testing for the most common genetic variant (ΔF508)[5] involved in the development of cystic fibrosis. The results of such tests pertain only to the sequences those probes could detect (a gene for cystic fibrosis, in our example) and could only be considered as targeted or primary to the initial purpose of the testing. However, as mentioned in the Plan, most diseases are not monogenetic, as cystic fibrosis is (and even in cystic fibrosis ΔF508 accounts for about 66–70 % of cases,

---

[1]Screening conducted on reproductive cells (egg cells and spermatozoids) prior to in vitro fertilisation in order to choose those not carrying variants implicated in disease developments.

[2]Testing conducted on an entire population or parts of the population as part of preventive and early diagnosis health interventions.

[3]In rare diseases, the number of individuals suffering from the condition in question is so small and the variability among those people consequently so large and that it is difficult to construct the list of the most common symptoms and potential treatments. Characterisation on the molecular level, however, could provide a more stable source of reference for treatment and for diagnostic/prognostic purposes.

[4]Many diseases, and cancer in particular, show individual differences in response to currently available treatments. This suggested tailoring the treatment to the individuals by looking at the differences they exhibit in symptoms but more recently on molecular level, and in particular in their genomic sequences.

[5]ΔF508 is a deletion (indicated by Greek the letter Δ) of three nucleotides that results in a loss of the amino acid phenylalanine (F) at the 508th position on the protein.

leaving more than 30 % of cases to be looked for outside this genetic spot). Technological developments enabled probing of an entire genome in looking for underlying variants that are linked to the development of multifactorial, complex diseases (a genome-scale testing). This, in turn, opened a possibility of discovering variants implicated in the development of other diseases. As these variants were not the initial aim of the test, they were termed as incidental,[6] or secondary findings. In addition to being a practical issue to be addressed,[7] incidental finings raised several ethical issues that require management.

Unsurprisingly, the transition from targeted genetic testing to genome-scale testing first occurred in the research context as most of the techniques it involved were first used for exploratory, that is, for research purposes. For this reason, the literature on incidental findings mostly considers the implications and the management of incidental findings in research setting (see, e.g. Wolf et al. 2012; Knoppers et al. 2013).

However, it is worth exploring its implication in the clinical setting as the whole exome[8] and genome sequencing are rapidly entering the clinic (see, Green et al. 2012; Berg et al. 2011). In what follows, we will focus on incidental findings in the clinic and we refer the readers interested in incidental findings in the research setting to the literature that covers this area.

2 Reporting of Incidental Findings in the Clinic

The genome-scale testing can uncover genetic variants that have a potential to cause, but which have not yet resulted in clinical conditions. The debate on incidental findings is a complex one (Blackburn et al. 2015), but the core of it often revolves around two main aspects: physicians' duties to inform and patients' rights to be informed, as well as costs/benefits of preventive actions on both individual and population health level. As this book is concerned with Ethical Counselling for Patients and physicians, we will here focus only on the debated surrounding physicians' duties to inform and patients' rights to be informed in relation to incidental findings.

---

[6]We would like to note that incidental findings did not originate with genomics. They first appeared in radiology with radiographic imaging as technology that made it possible to "see", that is, to detect previously unsuspected health conditions.

[7]It could be worth noting that there is also a debate regarding incidental findings that questions the utility of the concept itself, because genome-scale screening itself includes collecting all the variants in one patient; therefore, it is inevitable to discover variants that are unrelated to the original purpose of the test. A more suitable term, according to these authors (see, Allyse and Michie 2013), is 'secondary findings'. However, as the concept has both historic and analytic utility, it deserves consideration in a handbook of ethical counselling.

[8]Exome refers to a part of the genome (genomic sequence) that is expressed in cells/tissues tested —genome is always the same, no matter what tissue/cell or at what point in life, while exome varies from tissue to tissue and from cell to cell and also in time.

Some of the most pressing ethical questions in the debate on the management of incidental findings are as follows:

- should the physician be obligated to report all such findings back to the patients, or just some findings—in that case, which ones, or none?
- should the patients have a right to demand such results to be delivered to them under all circumstances, or should they be allowed to refuse to receive any such information?
- should a patient with a genetic variant implicated in the development of serious, but preventable/treatable clinical condition be allowed to refuse to know such information and consequently withhold it from family members that can also be carriers of that same genetic variant?
- should some genetic variants that can cause preventable/treatable clinical conditions that come up as incidental results in genome-scale screening testing as be actively sought in such testing, becoming thus a secondary instead of incidental finding, or, in fact, a regular finding of the clinical screening?

When discussing the duty on the side of the physician to report unanticipated results of medical tests in molecular and genomic medicine, the attention is mostly given to determine their clinical utility[9] and whether the treatment for the corresponding clinical condition is available. As discussed in several chapters, different health conditions have different underlying genetic mechanisms and hence different predictive powers when genetic testing is performed.

Some genetic variants have 100 % penetrance, like in the cases of many monogenetic dominant traits. Others have very complex probabilistic character, and their development depends on various factors besides the underlying genetic sequences, such as social and geographical circumstances and personal lifestyles. Correspondingly, being a carrier of a genetic variant for monogenetic dominant disease has significantly different implications than being a carrier of a genetic variant that implies merely an increased risk for a development of a multifactorial disease sometime in the future. In case of a dominant monogenetic disease, the onset of the disease is undisputable. When will the onset begin and how severe the clinical image will be is the only uncertainty that exists with respect to these conditions. Knowing that someone is a carrier of such variant can result in starting the management of the disease early on or opting for prenatal genetic testing. However, learning that someone is a susceptibility carrier means either that the

---

[9]By *clinical utility*, we intend that the results clearly identify and/or indicate targets of clinical action. For example, a result that shows the presence of ΔF508 has high clinical utility, as it detects a variant known to be causing cystic fibrosis and suggests treatment options. Contrary to this, a results that shows the presence of genetic variant thus far only shown to be loosely related to a condition in animal models (rats or mice, for example) is of low clinical utility because even though it can strengthen a hypothesis that it is indeed important in that condition and support further research, it does not currently offer prognostic, diagnostic or treatment options.

disease can be prevented by early intervention into patients' environment and lifestyle or that the patient can be exposed to unnecessary stress about a condition that may never develop. For example, such carriers may be advised to follow a strict dietary and exercise regime that can affect their lives in a positive way—a patient's general health conditions improves significantly due to a change in regime, but also in a very negative way—a patient's need to follow a schedule interfering with some other aspects of his/her life can cause her unhappiness and actually contribute to poorer rather than improved health by itself (and all because of some clinical condition that may or may not ever develop). On the other hand, despite all the technological developments and the amount of knowledge produced in recent years on human biology, there are diseases and health conditions that remain untreatable by currently available methods. This is the case not only with complex and chronic diseases, but also with some monogenetic dominant conditions. One such example is Huntington's disease for which, despite the gene and the mechanism of development being fairly understood, no therapeutic options is yet available.

All of these factors should be taken into consideration when deciding about reporting incidental findings. A recent attempt to provide a legal ground on which reporting of incidental findings should take place came from the American College for Medical Genetics and Genomics in form of a policy statement on clinical sequencing followed by the *Recommendations for Reporting of Incidental Findings in Clinical Exome and Genome Sequencing* (see, Green et al. 2013). This document considers variant frequency, penetrance of genes involved, the strength of association between specific gene abnormalities and the condition, and the potential for medical intervention to mitigate disease as the factors to be taken into account when making decisions about reporting incidental findings. However, the document goes a step further to present an argument for a professional duty to not just report incidental findings to patients, but to actively seek for some genetic variants whenever clinical screening is performed. Based on the factors mentioned above, the document then proposes a list of conditions that should be sought after in every case of clinical screening. Although criticised by some as representing an attack to patient's autonomy (see, Wolf et al. 2013), and as lacking actual legal power, the document has also been praised as "an initial attempt to set a professional standard for best laboratory practices" (see, McGuire et al. 2013). The list of conditions proposed by the American College of Medical Genetics and Genomics can certainly serve as valuable tool in guiding the decision-making in the cases of incidental findings in clinical sequencing. But, although the document itself states that the list is a not at all conclusive and is a developing one, there are many situations in the current clinical practice that fall out of the scope of this document. Personalised Ethical Counselling can provide a valuable resource for case-by-case-based clinical decision-making on these matters.

3 Personalised Ethical Counselling and Incidental Findings

As proposed in this guide, a patient-centred approach through dual methodology facilitates trust and provides mutually beneficial dialogue with better outcome for the patient. We argue this because offering a more equal ground in the decision-making process can convey a message on the physician's side that not only the patient's clinical image is of importance to him, but also patient's Personal Philosophy and the life he/she finds acceptable. This recognition, in return, can make the patient more open to discuss and take into consideration options previously considered incompatible with his or her Personal Philosophy and can aid in making a truly informed and aware clinical choice. The prerequisites for reaching an ethically legitimate clinical decision, therefore, fall on both sides of the doctor–patient relationship. Incidental findings are, in other words, the perfect testing ground for both of the methodologies introduced by this book, namely the one for professionals and the other one for patients. On the side of medical practitioners, a common recognition that incidental findings may arise from medical tests and a common awareness about the differences in the implications of different types of such findings for the patient and his/her family are one part of the equation. Within this context, the physician may be required to go beyond his/her role as the professional figure that is expected to cure patients, in order to become (as argued in chapter "Ethical Counselling for Physicians") a partner for both therapeutic and non-therapeutic decisions in care pathways. The methodology for Ethical Counselling directed at professionals we introduced in this book may also be a viable means for the achievement of such goal and for preparing the physician to deal with the ethical dilemma behind the case in partnership with the patient.

The other complementary part is that the patients undergoing such testing should themselves understand that such findings might and probably will arise and the kind of implications these findings can have for themselves and their family members (a potential to reveal not yet developed but treatable condition, not yet developed but untreatable condition, increased risk for treatable or untreatable condition, and that other family members can be affected as well). Ethical Counselling plays an important role within this context, in supporting and guiding the patient towards the identification of values to be promoted, and choices to be made, on the basis of his/her Personal Philosophy. Whether the ethical dilemma deals with the exercise of patient's right-not-to-know (see chapter "The 'Right-not-to-Know'"), or with a specific reproductive choice (see chapter "Genetic Testing and Reproductive Choices"), or with the decision as to be informed about particular incidental findings arising from the performance of the test (as discussed in this chapter), a number of potential issues (if not dilemmas) may come into patient's clinical path, which may profit from the support ethical counsellors and medical professionals.

In the case of Julie, the clinical implications of the incidental finding are mild. On the other hand, the condition can, in some rare cases, cause serious clinical outcome, which can be mitigated or even prevented by a treatment at early stages of development. A strictly professional/clinical decision on the physician's side would be to inform Julie about such finding and about the health consequences as well as treatment and lifestyle intervention options that are at their disposal. This is precisely what the physician in Julie's case did. However, such information can have a considerable effect on Julie's life and that of her children and family in general. Julie's choice not to know which one of her children is under risk should play an important factor in the ethically legitimate decision-making in the clinic. She does not want to make conscious or unconscious bias in raising her children as equal. A thought experiment in this case could be that, rather than twins, Julie is carrying a single child and receives the same karyotype results: her unborn is XYY. In this scenario, the fear of potential unequal treatment is not present. If Julie is aware of possibility to receive results about minor health conditions (like XYY) along the results of more serious conditions (like Down syndrome), she is able to consider what her preferences (yes to all results, or no to some results) with respect to this are prior to receiving the actual results. If doctors are aware that conflicts can occur between patients' Personal Philosophy and medical advice, they are able to consider the possibility of such conflicts in deciding whether and how to communicate the results back to Julie. The final output, thus, is more informed patient choice and more patient-centred medical advice that together bring the best possible medical decision.

References

Allyse M, Michie M (2013) Not-so-incidental findings: the ACMG recommendations on the reporting of incidental findings in clinical whole genome and whole exome sequencing. Trends Biotechnol 31:439–441

Berg JS, Khoury MJ, Evans JP (2011) Deploying whole genome sequencing in clinical practice and public health: meeting the challenge one bin at a time. Genet Med 13:499–504

Biesecker LG (2012) Opportunities and challenges for the integration of massively parallel genomic sequencing into clinical practice: lessons from the ClinSeq project. Genet Med 14:393–398

Blackburn et al (2015) Management of incidental findings in the era of next-generation sequencing. Curr Genomics 16:159–174

Heather L et al (2015) Management of incidental findings in the era of next-generation sequencing. Curr Genomics 16:159–174

Green RC, Berg JS, Grody WW et al (2013) ACMG recommendations for reporting of incidental findings in clinical exome and genome sequencing. Genet Med Official J Am Coll Med Genet 15:565–574

Green RC et al (2012) Exploring concordance and discordance for return of incidental findings from clinical sequencing. Genet Med 14:405–410

Knoppers BM et al (2013) Populations studies: return of research results and incidental findings policy statement. Eur J Hum Genet 21:245–247

McGuire AL et al (2013) Ethics and genomic incidental findings. Science 340:1047–1048

Wolf SM, Annas GJ, Elias S (2013) Patient autonomy and incidental findings in clinical genomics. Science 340:1049–1050

Wolf SM et al (2012) Managing incidental findings and research results in genomic research involving biobanks and archived datasets. Genet Med 14:361–384

Oncofertility

Alma Linkeviciute and Fedro Alessandro Peccatori

Abstract This chapter tackles ethical issues surrounding fertility preservation in cancer patients by offering an overview of the main ethical problems which might arise in patients. Main arguments supporting and rejecting the fertility preservation practice are presented including fertility preservation for young children, reproductive decisions faced by cancer survivors and cancer treatment during pregnancy. Moreover, it also shows that decisions to preserve fertility might require further decisions in the future about the use of cryopreserved gametes.

Keywords Oncofertility · Fertility preservation · Cancer patients · Cancer during pregnancy · Reproductive ethics

In the vast majority of cases, cancer patients are diagnosed in advanced age, but sometimes, this disease also affects young adults, adolescents and children. Even though the primary goal of cancer treatment remains to combat the disease, there are a number of other considerations to be taken into account. Fertility preservation is especially relevant for patients of reproductive age and children. More sensitive diagnostic tools and advanced therapies have contributed to earlier detection, more effective treatment and higher cancer survival rates. Despite existing controversies (see chapter "The Centrality of Probability" and chapter "Overdiagnosis"), it increasingly becomes reasonable and relevant to think about the quality of life after cancer. Therefore, over the last decade an interdisciplinary field of oncofertility evolved advocating the importance of informing and counselling cancer patients (and their relatives, when appropriate) about fertility preservation options, should

A. Linkeviciute (✉)
Dipartimento di Scienze della Salute, University of Milano, Milan, Italy
e-mail: alma.linkeviciute@ieo.eu

A. Linkeviciute
Department of Experimental Oncology, European Institute of Oncology (IEO), Milan, Italy

F.A. Peccatori
Fertility and Procreation Unit, Gynecologic Oncology Division, European Institute of Oncology (IEO), Milan, Italy
e-mail: fedro.peccatori@ieo.it

© The Author(s) 2016   97
G. Boniolo and V. Sanchini (eds.), *Ethical Counselling and Medical Decision-Making in the Era of Personalised Medicine*, SpringerBriefs on Ethical and Legal Issues in Biomedicine and Technology, DOI 10.1007/978-3-319-27690-8_10

cancer treatments endanger their future reproductive choices. Preserving fertility before cancer treatment means that patients are provided with a future opportunity to overcome treatment-induced infertility. However, it is important to note that none of fertility preservation options, available to both male and female, can guarantee future ability to parent a child. Moreover, different modes of fertility preservation have varying acceptance levels by the individual patients due to the state of establishment and efficacy, mode how it is performed and a general prognosis the patient has. Fertility preservation also has some wider implications that ideally have to be thought of before opting for or against the procedure.

It has been shown and argued extensively that fertility preservation is of extreme importance for most young cancer patients and by now it is considered to be a standard, or at least highly recommended, practice to inform all cancer patients about possible cancer treatment effects on their future fertility (Linkeviciute et al. 2014). However, patient-centred personalised care does not end with providing a patient with certain information. Some patients might need more detailed explanation what information about fertility preservation means in their case and require some time to consider their options and make a well-informed choice. This is where communication (see chapter "Reasons and Emotions") and interdisciplinary collaboration (Woodruff et al. 2014) play a crucial role.

1 Fertility Preservation in Oncology

There are different fertility preservation options for men and women. Some of them, like sperm cryopreservation for men and embryo and oocytes cryopreservation for women, are now considered as established and clinically approved methods to preserve fertility before starting cancer treatment. Sperm cryopreservation for men is a simple and relatively inexpensive procedure that has been used for many years in assisted reproduction. Traditional recommendation is to bank at least three samples of semen that are obtained with 48-hours abstinence between samples. If treatment has to be started immediately, fewer samples can be banked. Embryo cryopreservation is the most established fertility preservation method for female cancer patients. The downside of this method is that it also requires a source of sperm, which can be a problem for adolescent girls and single women who do not wish to use the donor sperm. In addition, not all patients feel comfortable with freezing embryos due to special moral value attributed to them.[1] Therefore, these patients might favour the option of oocyte cryopreservation. Both methods might require cancer treatment delay by at least 2–3 weeks in order to perform oocyte retrieval (Shan et al. 2011). It has also to be noted that to increase the chances for future pregnancy, the higher number of cryopreserved oocytes might be better and this would require two or three ovarian stimulation cycles that might not be compatible with optimal cancer treatment.

---

[1]This specific point is tackled in more detail by Robertson (2014) and Lauritzen and Vicini (2011).

Other methods, such as surgical transposition of ovaries outside irradiation field and ovarian suppression, can be used for reproductive age women and adolescents who have reached puberty. Concerning the children who have not reached puberty yet, only experimental fertility preservation methods are available, that is testicular tissue cryopreservation for boys and ovarian tissue cryopreservation for girls (Shan et al. 2011; Linkeviciute et al. 2014). The major concern surrounding fertility preservation for cancer patients is that, in the case in which patients opt for fertility-sparing procedures, their cancer treatment might have to be delayed and delay can vary between a few days and a couple of months. Another concern is the possibility to "reseed" cancer when performing autologous transplantation of previously frozen testicular or ovarian tissues. Moreover, it can also be argued that not all cancer patients have good prognosis and they might never be fit enough or even survive to make the use of cryopreserved gametes, tissues or embryos. As we will see in the following section, fertility preservation can serve as a therapeutic intervention in itself since it gives patients hope for overcoming the disease.[2] However, counselling about contraception is also very important and should be an inclusive part of oncology care (Han et al. 2015).

2 Ethical Debate Surrounding Fertility and Cancer

2.1 Arguments Supporting Fertility Preservation for Cancer Patients

Overall, the liberty to make one's own reproductive choices freely is considered to be one amongst the most fundamental human rights. Therefore, if someone's fertility is endangered by the disease, it would be reasonable to make some attempts to preserve it. This is one of the major arguments supporting the parental choice to pay for their children's fertility preservation which has the potential to ensure the open future where children affected by cancer and exposed to fertility impairing treatments will be able to have genetically related children of their own (Quinn et al. 2012). This line of reasoning is very well illustrated by Claudio's case (Case 2 in "The Plan") whose parents chose to pay for sperm cryopreservation presuming that their son might want to have children in the future. The second argument refers to the duty of not harming, or, when harm is unavoidable, to offer the repair for the damage. Offering fertility preservation before cancer treatment can serve as reproductive insurance for cancer patients who would like to consider parenthood in the future. The future prospects of building a family is also presented as an argument favouring fertility preservation for cancer patients since it can provide patient with hope that there can be life with and after the disease (Pennings and

---

[2]Patient attitudes and concerns surrounding the parenthood after cancer are explored in detail by Goncalves et al. (2014).

Mertes 2012). Nonetheless, as we can see in Claudio's case, the decision to build a family might not be as simple as initially presumed. Decision to embark with parental project includes at least two people or more, when gamete donors and/or gestational carriers are required. The latter two are ethically problematic occurrences in themselves, even if in reproductive choices women are usually seen as privileged decision-makers. Claudio is taking into consideration that the suggested procedure of "intra-cytoplasmic sperm injection" (henceforth ICSI)[3] involves a number of medical procedures with risks and possible toxicities for Francesca. Considering that success rates are roughly about 1 in 3 having second thoughts about the procedure can be rather common.

2.2 Arguments Objecting Fertility Preservation for Cancer Patients

As it has already been mentioned, fertility preservation procedures might be delaying cancer treatment, exposing the patient to additional risks such as surgery and/or ovarian stimulation, and in addition, offering patients the false hope, especially when only experimental fertility preservation methods are available. In Claudio's case, it has not delayed the treatment for very long and we can see that he is healthy and cancer free now. He also had his fertility preserved. However, even established fertility preservation methods offer rather low success rates which can highly depend on the individual circumstances. Therefore, preserving fertility might not always be in the best interest of the patient, considering additional risks and low rates of future success in conceiving a child. Concerns about the welfare of the future children are also at the heart of the debate. One worry is that children of cancer survivors might inherit the disease, and even though such a risk can be reduced by using PGD techniques,[4] it can still raise uncertainties. For instance, the only viable embryo might be the one with mutated gene, which causes cancer early in life. In such case, a new choice will have to be made—to have a child who is likely to suffer from cancer or to have no genetically related children at all. Another worry concerning the future children is the health and lifespan of the parent (or both parents) who have been affected by cancer. It is possible that cancer survivors might be of poorer health and have less energy to look after their children. It can also be argued that witnessing parental battle with cancer can be traumatic for a child. Moreover, cancer survivors might die prematurely leaving their children orphans before the children are capable of taking care of themselves. Here, the arguments

---

[3]ICSI is a mechanical insertion of a sperm into an oocyte as the membranes of gametes are affected by the cryopreservation and thawing procedures, which reduces the chances of fertilisation using a standard IVF procedure.

[4]PGD refers to pre-implantation genetic diagnosis. It is a technique used to identify embryos with faulty genes and/or chromosomes which are known to cause birth defects or increase predisposition to certain diseases such as cancer.

over absolute and reasonable welfare clash as most probably nobody is born into ideal circumstances and protected of potential loss or illness of their parents. However, Claudio can still wonder if having a child is a right thing to do when he knows that he might have health problems early in life compared to other people. In addition, some might have reservations towards the use of assisted reproduction technologies (henceforth ART) based on religious convictions and/or social and cultural background affecting their views on life. This might be relevant to those who are still considering fertility preservation and also those who already have their gametes preserved, especially if their gametes were preserved when they were children as ways how people see life can change over their lifetime.

Furthermore, it can also be expensive to carry out fertility preservation and other ART procedures, and it can be argued that it is not the best way to allocate the scare resources, be it public or personal funds. For instance, Claudio's parents might have had to borrow money to afford his sperm freezing or if it was taken from their savings it could have been used for Claudio's education, other healthcare needs, etc. It has to be recognised that in some cases, fertility preservation is simply a fee for a false hope; in other words, it is a parental wish to become grandparents themselves presuming that this is what their child would want. Promoting fertility preservation for all cancer patients might have wider social implications such as reinforcing the belief that parenthood is an essential feature of meaningful life and genetic parenthood is superior to other ways to form a family (Pennings and Mertes 2012). Hence, sometimes preserving fertility can also have a therapeutic value encouraging patients to stay more optimistic, believing that there is a bright future ahead and that they have a possibility to resume healthy and normal life after their disease is cured. However, each situation is different and not every cancer survivor might see parenthood as an optimal choice to be made.

2.3 Considerations Concerning Future Fertility of Children Affected by Cancer

Adolescents who have reached puberty can benefit from the same fertility preservation methods available to adult cancer patients. Prepubertal children, however, can only be offered experimental interventions that are still considered research. Fertility preservation interventions, even if experimental, are often argued to be justifiable if the risks associated with a procedure are minimal and the patient is likely to benefit from the procedure (Pennings and Mertes 2012). It is advised to include the older children and adolescents in decision-making process by providing a specialised support for the families by ensuring that all members of the family understand the purpose and nature of the intervention (Woodruff et al. 2014). Some aspects of fertility preservation for children and adolescents still remain controversial, especially in cases where parents made financial, emotional and physical effort to preserve their children's future fertility (Quinn et al. 2012) as open future argument is not limited to the right to be a parent but also includes the right to refuse parenthood. As Claudio's case illustrates, the right to refuse parenthood might be complicated if there are expectations to pursue parenthood from the parties who contributed with an effort to make this choice available. Although there is no guarantee that using previously cryopreserved sperm will result in successful pregnancy and healthy birth, Claudio and Francesca might face a pressing expectation from Claudio's parents to try at least once.

2.4   Concerning Cancer Treatment During Pregnancy

Cancer complicates about 1 in 1000 pregnancies. It has also been reported that, due to delayed childbearing and other factors, the number of pregnancies complicated by cancer is rising (Oduncu et al. 2003). When a pregnant woman is diagnosed with cancer, a conflict between maternal well-being and foetal benefit might occur. The scale of the conflict usually depends on the type of cancer, treatments available and the stage of pregnancy at the time of diagnosis. In early pregnancy, optimal therapy for a mother usually compromises foetal well-being, while during the second and third trimester, adapted treatments are almost routinely given to pregnant patients. Initial safety data are reassuring, and there is no scientific evidence that terminating the pregnancy would increase the chances of survival or quality of life for a patient (Peccatori et al. 2015). It has to be acknowledged, however, that there are a number of other factors that play a role when taking treatment decisions in situations like Anna's.

For example, in Anna's case (Case 3 in "The Plan"), there are three therapeutic paths that can be followed. The first is standard treatment, which is given to non-pregnant women with Anna's condition and constitutes of surgery to remove the tumour followed by standard chemotherapy, radiotherapy and hormonal therapy for 5 years. Unfortunately, it is not compatible with pregnancy because it compromises the development of the foetus. Therefore, if Anna choses the standard treatment, the very much wanted pregnancy will have to be interrupted. The second option available to Anna is adapted treatment, which starts with surgery to remove the tumour and is followed by adapted chemotherapy, which is, in turn, compatible with foetal development. Hence, choosing the second option would pose some risks for Anna and also the foetus. It is known that adapted treatment might have lower response rates. There are also some known risks to the foetus such as preterm delivery and lower birth weight, and other risks such as toxic effect on cognitive and cardio functions as well as future fertility are still theoretical. If Anna chooses this option, the radiotherapy and hormonal therapy will be delayed until after the delivery of the baby. One more possibility is just surgery to remove the tumour during the course of pregnancy, postponing all other treatments until after the delivery of the baby. Breast surgery is generally considered safe and compatible with pregnancy, but Anna will not be protected from the potential cancer spreading.

As we can see, none of the choices is optimal for Anna and the foetus. Moreover, Anna is not alone in the decision-making process. She is also not the only one to bear the implications her choice might have. We also saw that her husband Matteo is against any choice that could compromise Anna's survival. The decision they have to make might be directed by the perception of the foetus in a light of personal moral values but also by very practical considerations like who will take care of their son and a new baby if Anna is not well enough to look after them or dies if her cancer progresses.

Usually, the respect for maternal autonomy and the balance between potential harms and expected benefits are held to be of paramount importance when approaching dilemmas arising in cancer and pregnancy care (Oduncu et al. 2003). It also includes, but is not limited to, the welfare of the future child as well as existing children (when present), support available and general prognosis for the mother. Pregnancy and cancer diagnosis can both be very emotional events where taming the emotions (see chapter "Reasons and Emotions") might have to be an initial step before exploring patient's Personal Philosophy and views about life.

3   Personalising Ethical Counselling for Oncofertility Patients

In addition to helping a patient to look at his or her cancer and its impact on the future life, Ethical Counselling could also serve as a *reflection slot*, where patients can reflect on personal expectations towards fertility preservation, intended or current pregnancy. Success rates, risks and benefits, the costs and future implications of fertility preservation can be overwhelming and complex information. There have been reports that even some doctors feel that they lack knowledge concerning fertility preservation for cancer patients (Linkeviciute et al. 2014). Exploring which way of action is the most compatible with one's values and beliefs can be even more complex.

Fertility preservation, just as illustrated by Claudio's case, is not "one-stop shop". Once gametes, embryos or tissues are cryopreserved, they can be thawed and used at any time. Therefore, it is very important to decide and document the patient's wishes from the very beginning. Who can use the gametes, tissues and embryos, for what purposes (self-procreation, donation to third party procreation, research) and under what circumstances this biological material can or should be discarded? What type, if any, of posthumous use of gametes is acceptable for a patient? Moreover, ethical uncertainties might arise years later after fertility has been preserved. In addition to Claudio's concerns, one might wonder how acceptable is the use of PGD or the use of surrogacy and who could serve as a gestational carrier. Could Claudio's parents use Claudio's sperm to have a genetically related grandchild if Claudio dies without using his sperm himself? Anna's case is even more dramatic as it requires balancing the risk for the foetus and ensuring the best possible treatment outcomes for her. In such cases, the collaboration among the multidisciplinary team will be necessary as it is complicated not only from a medical point of view but also can put a lot of emotional and existential strain on the patients, their relatives and also clinicians.

References

Goncalves V et al (2014) Childbearing attitudes and decisions of young breast cancer survivors: a systematic review. Hum Reprod Update 20(2):279–292

Han SN, Van Peer S, Peccatori FA, Grizi MM, Amant F (2015) Contraception is as important as fertility preservation in young women with cancer. Lancet 385(9967):508

Lauritzen P, Vicini A (2011) Oncofertility and the boundaries of moral reflection. Theol Stud 72 (45):116–130

Linkeviciute A, Boniolo G, Chiavari L, Peccatori FA (2014) Fertility preservation in cancer patients: the global framework. Cancer Treat Rev 40:1019–1027

Oduncu FS, Kimming R, Hepp H, Emmerich B (2003) Cancer in pregnancy: maternal-foetal conflict. J Cancer Res Clin Oncol 129:133–146

Peccatori FA, Corrado G, Fumagalli M (2015) After gestational chemotherapy, the kids are all right. Nat Rev Clin Oncol 12(5):254–255

Pennings G, Mertes H (2012) Ethical issues in infertility treatments. Best Pract Res Clin Obstet Gynaecol 26:853–863

Quinn GP, Stearsman DK, Campo-Engelstein L, Murpphy D (2012) Preserving the right to future children: an ethical case analysis. Am J Bioeth 12(6):38–43

Robertson JA (2014) Egg freezing and egg banking: empowerment and alienation in assisted reproduction. J Law Biosci 1(2):113–136

Shan DK, Goldman E, Fisseha S (2011) Medical, ethical and legal considerations in fertility preservation. Int J Gynaecol Obstet 115:11–15

Woodruff TK, Clayman ML, Waimey KE (2014) Oncofertility communication: sharing information and building relationships across disciplines. Springer, New York

Overdiagnosis

Giulia Ferretti

Abstract Screenings appear to be one of the most promising approaches to tackle cancer based on the assumption that prevention and proactive management of risky lesions is the best strategy to reduce fatalities from invasive cancers. However, the opportunity of early detection implies the possibility of unwanted potentially harmful outcomes, such as *false-positive results* and *overdiagnosis*. We argue that the balance between possible benefits and harms has to be established by patients deciding whether to undergo screening. Moreover, we propose Ethical Counselling as a tool for positively coping with these questions.

Keywords Overdiagnosis · False-positive · Cancer screening · Decision-making · Consent

Thanks to the advancements in screening technology and treatments for cancer patients, many promising approaches have changed the impact of cancer on patients' health dramatically for the better and have significantly reduced mortality.[1] However, notwithstanding major improvements in cancer prognosis at all stages of disease, still little doubt exists as to the fact that prevention and proactive management of risky lesions is the best strategy to reduce fatalities from invasive cancers. Screening technologies currently aim at detecting benign lesions and in situ

---

[1]*Mortality* refers to the number of people who died within a population and differs from morbidity, which refers to the state of being diseased or unhealthy within a population. Among the innovations that have contributed to reduce cancer mortality, we might find: chemotherapy, adjuvant therapy, combination chemotherapy, hormone therapy (for breast and prostate cancers), improvements in surgery techniques, radiation, immunotherapy (for breast cancer and lymphoma), targeted therapy (growth signals inhibitors, angiogenesis inhibitors and apoptosis-inducing drugs).

G. Ferretti (✉)
Dipartimento di Scienze della Salute, University of Milano, Milan, Italy
e-mail: giulia.ferretti@ieo.eu

G. Ferretti
Department of Experimental Oncology, European Institute of Oncology (IEO), Milan, Italy

© The Author(s) 2016
G. Boniolo and V. Sanchini (eds.), *Ethical Counselling and Medical Decision-Making in the Era of Personalised Medicine*, SpringerBriefs on Ethical and Legal Issues in Biomedicine and Technology, DOI 10.1007/978-3-319-27690-8_11

carcinomas that are associated with an increased risk of invasive cancer development.[2] The recognition and early treatment of such high-risk and/or pre-invasive lesions are considered able to help preventing progression of invasive diseases and, thus, to reduce the overall cancer incidence and mortality at a population level.

By reducing cancer incidence and mortality, cancer screenings are proven beneficial on population health level, and therefore, in the past few decades, participation to screening programmes has been widely suggested to some risk categories of patients (Welch et al. 2011). However, the effectiveness and safety of some screenings are controversial matters, either because evidence is lacking as to determine whether screening reduces mortality, or because the balance between benefits and harms is uncertain or challenging to weigh.

A major limitation of screening tools is their sensitivity (i.e., the probability to test positive given the presence of the disease, as seen in chapter "The Centrality of Probability"), which is still open to improvements, and the defect in their capacity to discriminate between early-stage cancers requiring treatment and those that do not. Indeed, the opportunity of early detection goes hand in hand with the possibility of unwanted potentially harmful outcomes for some individuals, such as *false-positive results* and *overdiagnosis*. The former represents a diagnostic mistake that can be rectified after additional tests, while the latter occurs when benign or slow-growing cancers are detected and subsequently treated.

Current cancer risk-management strategies are based on population-level risks, resulting in benefits for a minority at great risk and in harms for those individuals who will encounter *false-positive* result and *overdiagnosis*. The ethical debate on cancer screenings is controversial and complex, but it often revolves around two main themes: costs and benefits of *early detection* strategies on both individual and population health level, and patients' right to be informed vis à vis physicians' duty to inform. In this chapter, we will discuss cancer screenings from the clinical ethics standpoint focusing on decision-making based on statistical information, taking into account issues such as respect for autonomy, consent, and non-maleficence. The public health debate is left aside and for those interested we refer to the current literature on this issue.[3]

---

[2]*Screenings* can be defined as programmes aimed at a population, no member of which is thought to be at greater risk and often indicates a heightened risk of the condition, which has to be confirmed through further diagnostic tests. By contrast (as shown in chapter "Incidental Findings "), *testing* refers to a procedure performed on an individual who has been identified as being at high risk and aims at confirming the tested-for condition. Examples of widespread cancer screening methods are as follows: mammography screening for breast cancer, PSA test for prostate cancer, and HPV test and PAP cytology test for cervical cancer.

[3]For the public health debate on cancer screening see Wright and Mueller (1995), Holland (2007), Juth and Munthe (2012).

1 The Patient's Side: False-Positive Results and Overdiagnosis

At the clinical level, we may relate the major problems occurring in cancer screening either to the limitations of screening tests, or to the uncertainty of estimating further development of high-risk or pre-invasive lesions and/or slow-growing cancers. Concerning the screening modality and its possible failures, when optimal *sensitivity* and *specificity* are lacking, the test might provide a certain amount of *false-positive* and *false-negative* results that can be statistically calculated (see chapter "The Centrality of Probability"). In the context of cancer screening, the most common burden related to errors in test is the occurrence of *false-positive* results that indicate the presence of the condition, but usually are proven mistaken after additional diagnostic interventions such as biopsy, radiation or even surgery. Nevertheless, even if finally ruled out, *false-positive* results bring up unnecessary stress, anxiety and health risks due to the subsequent diagnostic exposure to additional tests such as biopsy or surgery that inevitably expose patients to the potential complications entailed by any similar medical intervention (Welch et al. 2011).

Overdiagnosis could be classified as a consequence of the uncertainty in evaluating further potentially life-threatening development of lesions and cancer. In general, it occupies either as a consequence of an ill-defined concept of disease or abnormality, or as a result of performing more sensitive, imaging and screening tests, the latters being the major source of overdiagnosis in cancer. Indeed, cancer screenings might detect benign lesions that have no potential to grow and might even spontaneously regress; or they might detect very slow-growing cancers that would never cause any symptoms in patient's lifetime (Croswell et al. 2010). In other words, *overdiagnosis* refers to the detection of a tumour that fulfils the pathological criteria for cancer, but which does not constitute a substantial health hazard for the patient.[4] Notably, the inability to identify reliable predictors for the progression from detected high-risk lesions to invasive carcinoma prevents the individualisation of risk management, and often treatment is recommended to all patients in order to reduce the risks of recurrence and mortality. Therefore, in addition to the above-mentioned burden of false-positive test results, overdiagnosis exposes patients to lengthy and unnecessary additional diagnostic tests and treatments (*overtreatment*). As it is well known, every treatment has side effects that, by the side of patients, are typically overwhelmed by the gained benefits. However, overdiagnosed patients gain no benefit from the worthless therapeutic plan, and

---

[4]In the context of cancer screening, low-grade in situ carcinomas (the non-obligate precursor of infiltrating carcinoma) are likely to cause a number of cases where the diagnosis of malignancy is made, but the tumour may not prospectively be fatal nor clinically relevant for the patient. Although screen-detected carcinoma in situ is not a life-threatening condition, it is often treated the same way as invasive cancer due to the inability to predict with certainty on the screening images or results its further development.

they are only exposing themselves to the potential harms and side effects of treatment. Several studies have evaluated the main cancer screenings' safety and effectiveness uncovering alarming data: the reduction of death risk is bound to a significant risk of *overdiagnosis*.[5]

Notably, though, the only evidence of *overdiagnosis* comes from the statistical information showing a rise of screenings and incidence of the given disease in the setting of stable mortality rates (Welch et al. 2011). This means that it is possible to assess whether *overdiagnosis* occurs only at a population level, whereas it is extremely difficult, if not impossible at all, to assess whether it occurs at an individual level, and in fact patients will (most likely) never know if they were *overdiagnosed* and *overtreated*. Therefore, we can only estimate a statistical risk of encountering the potential harmful outcomes, and decisions have to be made under individual uncertainty. Cancer screenings represent a case where probabilistic information and medical risk communication shape the patient's decision-making prospect. Nevertheless, just as patients may choose between treatment options, they should also have the chance to consult individually with physicians and/or counsellors evaluating benefits and risks in order to make the best possible informed decision about whether to undergo cancer screenings.

2 Respecting Patients' Autonomy in Cancer Screening Decision-Making

Respect for autonomy, especially in the context of screening, is not just a matter of giving patients the opportunity to make any decision. It requires both of them being informed about benefits and risks of screening, and a recognition of the fundamental *evaluative dimension* in their assessment of benefits and risks (Plutynski 2012). Notably, the recognition that patient's own values and beliefs, socio-familiar situation, context and ideas about how to judge what is right for him/her and how consequently to assess risk should be the unavoidable basis of a medical non-directive approach. The accumulating literature on patient-centred care, to which this guide contributes with a specific methodology for ethical counselling, shows that this approach facilitates trust and provides mutual beneficial dialogue between the parties, thus producing better outcomes for the patient in a wide range of clinical settings and medical circumstances. A non-directive approach, as we will see in this section, seems particularly valuable also in the case of cancer screening, which entails complex probabilistic information and delicate decisions to be made

---

[5]The rate of *overdiagnosis* for different cancer types changes widely among studies. In this work, we take into account three studies: a 15 years' follow-up after the end of the Malmö mammographic screening trial revealed a mortality reduction of 20 % and a 24 % risk of *overdiagnosis*; for lung cancer, a 16 years' follow-up after the end of the Mayo trial revealed a mortality reduction of 13 % and a *overdiagnosis* risk of 51 %; for prostate cancer, the European Randomized Study of Prostate Cancer revealed a mortality reduction of 20 % and a *overdiagnosis* risk of 67 %.

under conditions of uncertainty. Indeed, the only information available concerns the likely benefits and side effects—namely the *risks*—of testing, as well as the possible positive and negative consequences of alternatives. As shown in chapter "The Centrality of Probability", psychologists have extensively studied our understanding of probability and found that we systematically tend to misinterpret statistical information. Such difficulties affect both patients and physicians, and thus improvements in understanding and communication of statistical information appear to be a prerequisite to enable patients making informed decisions. Therefore, clinicians should first manage understanding probabilistic data and then provide the patient with impartial and understandable information on screening effectiveness and safety in terms of possible benefits and risks.

Nonetheless, communicating probabilistic information about risks to patients is more difficult than it might be expected. It has been shown that individuals find very problematic understanding risk in probability and percentage format. For example, the common cognitive mistake of *base-rate fallacy* (see chapter "The Centrality of Probability"), often leads to an overestimation of screening sensitivity and effectiveness. At the same time, physicians should be careful in presenting statistical fallacies in order not to produce unintended consequences such as patients' underestimation of screening benefits and distrust in medicine. Lately, a growing set of the literature has indeed focused on the effects of presentation format of information for the sake of helping patients processing it and using it in their risk interpretation. A wide consensus rests on the idea that physicians should manage different statistic communication formats such as percentages, natural frequencies, base rates and proportions, absolute and relative risk reduction, cumulative probabilities, numerical and verbal probability information, graphs and risk ladders, in order to entail an adequate communication with patients by presenting them information in a personalised user-friendly format (Visschers et al. 2009).

As this guide advocates, when a particular clinical decision is reasonable and represents an individual's embodiment of his/her Personal Philosophy, the notions of medical benefits and harms cannot be disjoined from patient's judgment. Whether a particular risk is worth taking or not, it rests on a decision that is inescapably the result of life plans, goals, values and beliefs of individual Personal Philosophy. In addition, the specificities of cases of *overdiagnosis* require the establishment of a close relationship, characterised by confidence and trust, between the professional and patient. The patient's decision to take up the risk of not treating a lesion that *may* develop into cancer requires that the doctor and the patient agree upon a number of actions to be potentially taken to monitor and eventually treat the condition at stake. Among these options, it is worth highlighting the possibility that the patient may leave into the hands of the physician, and his professional advice, the decision about how far going in taking the risk of not treating a condition detected through screenings. This means that, deciding whether to undergo a screening test, and eventually acting upon it, might require shifting from an informed consent to a trust-based consent, which could be very suitable to tie up all of the agents and the factual, as well as evaluative dimensions involved in such a decision (Boniolo et al. 2012; Sanchini et al. 2015). Grounding consent on trust and reciprocity rather than on information implies, in fact, switching from the mere transfer of information to a proper communication, a suitable way of taking into account also those aspects that are relevant for patient's choice, such as clinical circumstances and patient's values, beliefs and interests. The ethical counsellor should therefore work in such direction for a proper guidance of patient's decision to undergo cancer screenings.

3   Case's Analysis: Veronica

Veronica's case (Case 5, "The Plan") constitutes a good example of complex clinical decision-making under uncertainty where no universal "best choice" seems to be definitively available. As shown above, in cancer screening contexts respect for autonomy requires the acknowledgment of patient's aware and reasoned choice as the best possible settlement. In the case at stake, there are two alternative paths. The first possible solution is to undergo a mammography screening accepting the risk of its possible harmful side effects, whereas the second is to refuse screening participation and instead schedule a clinical surveillance programme. Therefore, respect for autonomy requires physicians both to provide a proper risk communication and the making explicit of patient's Personal Philosophy, in order to list his/her moral values, thus providing moral reasons to wisely choose among clinical and/or preventive alternatives.

It is known that biases may occur when interpreting research data on screening tools and clinical information (see chapter "The Centrality of Probability"). In particular, three potential biases could arise when evaluating screening effectiveness: the *lead-time bias*, the *length bias* and the *healthy volunteer bias*.[6] These biases are relevant in the evaluation of screening programmes because they can significantly falsify the results and thus mislead both policy-makers in assessing health policy effectiveness and physicians in considering patients' participation benefits and risks. Thus, Veronica's physician primary deontological duty is to manage statistical information in order to make a professional evaluation based on the best possible evidence on mammography screening. As a result of a proper and adequate factual evaluation of breast cancer screening's safety and effectiveness, and in addition to individual patient's anamnesis and guidelines, the physician may end up judging the participation to screening as the right clinical option for the case at stake. Nonetheless, it is known that most patients tend to follow physicians recommendations when considering whether to participate breast cancer screening

---

[6]The *lead-time bias* refers to the added time of illness produced by the diagnosis of a condition during its pre-clinical phase. The *length bias* is typical of cancer screening because it refers to the inclusion in survey of more slow-growing cancers with longer disease duration and better prognosis than fast-growing cancers. The *healthy volunteer bias* concerns screening in which the participants are healthier than the general population showing spuriously increased benefits of the intervention.

(Metsch et al. 1998). As a consequence, doubts may still arise in considering the ethical legitimacy of recommending Veronica to undergo a mammography. Ethical counselling may help the physician clarifying the values at stake and providing reasons in favour of or against screening recommendation. For example, by embracing the principle of double effect (henceforth PDE),[7] the clinician can evaluate screening as ultimately beneficial. From the PDE standpoint, it is allowed suggesting and performing an action that the doctor foresees to produce a (possible) good and a (possible) bad effect. Notably, possible harmful outcomes such as *false-positive* results and *overdiagnosis* are not merely unwanted: they are unintended even if foreseen.

Besides physician's own evaluation of the screening safety and effectiveness, he should anyway acknowledge that the screening downsides would fall on the patient. Therefore, in order to properly respect Veronica's autonomy in decision-making, the physicians will have to provide her with some complete and understandable probabilistic information in a user-friendly format. In this case, for example, when discussing the probability of having breast cancer given a positive test results, the physician should ensure that a pitfall such as the *base-rate fallacy* would not prevent Veronica from understanding that mammography screening positive result represents a final cancer diagnosis only in 9 % of the cases, meaning only in 9 women over 98 who test positive (see chapter "The Centrality of Probability").

The proportionality between the disvalue of (possible) harmful outcomes and the value of the (possible) beneficial end should be assessed by Veronica who may have a personal view about what constitutes benefits and harms, and especially as to what extent risks are worth taking up or not. Ethical Counselling, by making explicit her set of beliefs and articulating her moral reasons would enable Veronica to clarify her Personal Philosophy, therefore helping her in deciding what to do. For example, she might consider the overall family well-being as an *end* in itself, and therefore, the guarantee of certainty with respect to her health condition comes clearly to be her primary responsibility towards her children and husband. Once she has set her *end*, she might still be unsure about which clinical *means* best conforms to it (see chapter "Reasons and Emotions"). Psychosocial outcomes such as anxiety, anticipated regret, perceived risk and worriedness are actually a matter of personal evaluation in balancing benefits and risks of screening and should be taken into serious account (Pace and Keating 2014). Therefore, Veronica might evaluate the early detection benefits as outweighing the possibility of unnecessary treatments and thus opt for mammography screening following the "better safe than sorry" approach. Conversely, she could prefer not to uptake the risk of being unnecessarily

---

[7]The principle of double effect (PDE) is often pleaded to justify the permissibility of an action that causes harm as side effect of promoting a good end. In medical ethics, PDE is often mentioned in discussions on palliative care, which aims at pain relief but may also cause respiratory failure. For the action being permissible, four conditions have to be verified at the same time: (i) the action's object is good or at least indifferent; (ii) the intended effect is the good and not the evil one; (iii) the good effect is not produced by means of the evil one; (iv) there is a proportionately reason for permitting the evil effect. For more information on the PDE see McIntyre (2014).

labelled with a frightful diagnosis or the necessity to undertake additional tests that are anxiety provoking and thus decide not to get screened and to schedule an annual breast clinical examination (BCE) or nothing at all. Of course, Veronica's emotions and expectations, in addition to her risk aversion, will play an important role in her decision-making process and cannot be totally dismissed nor should be passively followed. As a matter of fact, Ethical Counselling, by engaging in reason-giving, might aid Veronica in making a decision that really represents a way of achieving her moral *ends* taking also into account her psychosocial and emotional peculiarities.

4 Conclusion

A personalised Ethical Counselling may provide a valuable asset in guiding individual decision-making by addressing uncertainty and reflecting on the possible ways of action, their costs and consequences. As argued throughout this guide, patients have to be considered the privileged decision-makers when ethical issues are involved and therefore, Ethical Counselling could represent both for patients and physicians a space for reshaping and clarifying personal values and beliefs when facing complex clinical decisions. Indeed, the proposed methodologies recognise the importance of an adequate communication, the role of emotions and expectations in decision-making concerning potentially life-saving and harmful screenings, and understanding of probability. At the same time, a collective recognition on the side of physicians that cancer screenings might harm individual patients is one of the prerequisite for reaching an ethically legitimate clinical decision. Another prerequisite is that physicians acknowledge the need of a proper statistical knowledge. Correspondingly, patients deciding whether to undergo such screenings should be put in the condition of understanding probabilistic information in order to autonomously evaluate the balance between benefits and risks. Finally, all the parties should recognise and accept that the best choice for the patient depends on the importance and value he/she attributes to scientific uncertainty, potential benefits and harms of cancer screening. The ideal goal is to reach an informed choice about whether to undergo testing based on adequate clinical and statistical knowledge, and consistency between personal values, beliefs and choices. Indeed, as mentioned above, this situation may require the doctor and the patient to establish a trust-based relationship, which could be a way (i) to overcome some of the difficulties of providing patients with full and understandable information, (ii) to emphasise the central role of his/her Personal Philosophy when decisions under uncertainty are made and (iii) to minimise (e.g. through close monitoring) the chances that the (potentially) escalating disease significantly undermines patient's chances of healing and recovery.

References

Boniolo G, Di Fiore PP, Pece S (2012) Trusted consent and research biobanks: towards a "new alliance" between researchers and donors. Bioethics 26:93–100

Croswell JM, Ransohoff DF, Kramer BS (2010) Principles of cancer screening: lessons from history and study design issues. Semin Oncol 37:202–215

Holland S (2007) Public health ethics. Polity Press, Cambridge

Juth N, Munthe C (2012) The ethics of screening in health care and medicine: serving society or serving the patient? International library of ethics, law, and the new medicine. Springer, Netherlands

Mandelblatt JS et al (2009) Effects of mammography screening under different screening schedules: model estimates of potential benefits and harms. Clin Guide Ann Intern Med 151:738–747

McIntyre A (2014) Doctrine of double effect. In: Zalta EN (ed) The stanford encyclopedia of philosophy. Url: http://plato.stanford.edu/archives/win2014/entries/double-effect/

Metsch LR et al (1998) The role of physician as an information source on mammography. Cancer Pract 6:229–236

Pace LE, Keating NL (2014) A systematic assessment of benefits and risks to guide breast cancer screening decisions. JAMA 311:1327–1335

Plutynski A (2012) Ethical issues in cancer screening and prevention. J Med Philos 7:310–323

Puliti D et al (2012) Overdiagnosis in mammographic screening for breast cancer in Europe: a literature review. J Med Screen 19(Suppl 1):42–56

Sanchini V, Bonizzi G, Monturano M, Pece S, Viale G, Di Fiore PP, Boniolo G (2015) A trusted-based pact in research biobanks. From theory to practice. Bioethics (forthcoming)

Visschers VHM et al (2009) Probability information in risk communication: a review of the research literature. Risk Anal 2:267–287

Welch HG, Passow HJ (2014) Quantifying the benefits and harms of screening mammography. JAMA Intern Med 174:448–454

Welch HG, Schwartz L, Woloshin S (2011) Overdiagnosis. Making people sick in the pursuit of health. Bacon Press, Boston

Wright CJ, Mueller CB (1995) Screening mammography and public health policy: the need for perspective. Lancet 346:29–32

Conclusion: Choices

Giovanni Boniolo

Abstract Here we try to give a further perspective on what have been presented in the book, starting from what a choice is in the life of any human being.

This guide started with the presentation of five cases. These have constituted the "material" to illustrate two Ethical Counselling methodologies (one addressed to patients and the other to clinicians) and to explore some among the mostly relevant ethical questions that could intersect clinical decisions in the age of personalised medicine. Even if presented in an ideal-typic manner, those cases belong to reality; that is, they describe what real patients had to decide starting from ethically problematic situations.

Now the problems involved have been analysed and the patients have already chosen. But, which were the decisional scenarios? Did the patients and the physicians follow a well-structured methodology? It is precisely this last point that we consider central in any decisional process and that has been presented in the first part of this book: the idea that any decisional process should be reason-giving based. Indeed, even if the role and unavoidable component of emotions cannot and should not be excluded, what we tried to convey is that these choices, being existentially delicate and sources of positive and negative consequences, should be supported by reasons and matter of awareness by the decision-makers (be them patients or physicians).

Obviously, wrong choices can also be due to an improper communication or to a misunderstanding of the probability involved in the information concerning risk, as we discussed. However, a pivotal role is played by what we have called the

G. Boniolo (✉)
Dipartimento di Scienze Biomediche e Chirurgico Specialistiche, University of Ferrara, Ferrara, Italy
e-mail: giovanni.boniolo@unife.it

G. Boniolo
Institute for Advanced Study, Technische Universität München, Munich, Germany

© The Author(s) 2016

G. Boniolo and V. Sanchini (eds.), *Ethical Counselling and Medical Decision-Making in the Era of Personalised Medicine*, SpringerBriefs on Ethical and Legal Issues in Biomedicine and Technology, DOI 10.1007/978-3-319-27690-8

Personal Philosophy of those involved in the choice. Indeed, the two methodologies proposed in this guide, in addition to clarifying the situation from an ethical standpoint thus making patient's choice more aware, are also aimed at exploring and respecting the particularity of patients' Personal Philosophy.

Considering the decisional process, we wish to conclude recalling a couple of lessons coming from the Greek tragedies (see Williams 1993; Nussbaum 2001). To some extent, they could be thought of as a sort of compendium of most, if not all, the troubles that can affect the existence of an individual. They all rotate around a common aspect: the *tyche*. This is the "fortune" of an individual; that is, what can happen to him/her over his/her life independently of his/her will. The disease or the pathological situation affecting an individual and/or his/her family is exactly an event of this sort. The *tyche* of Giovanna (Case 1), Claudio (Case 2), Anna (Case 3), Julie (Case 4) and Veronica (Case 5) was a pathological event that generated diagnostic or therapeutic decisions involving ethical choices as well. However, how should we behave? What should be the best choice to be pursued?

Let us begin with Sophocles' *Antigone*. It narrates the choice taken by Creon— King of Thebes—after the end of a civil war which saw the fight and the death of two brothers: Eteocles—the former King of Thebes—and Polynices. Creon chose and ordered that Eteocles had honours but not Polynices, whose body had to remain unburied out of the city limits. Antigone, the sister of Eteocles and Polynices and in love with Haemon—Creon's son—, could not accept it. Thus she decided to infringe the edict and entombed Polynices. Consequently, Creon sentenced Antigone to be buried alive in a cave. Creon had to choose between the laws of the city and the piety for a sister's love towards her brother. Creon had no doubt: laws come first. Antigone had to choose between the laws of the city and the piety and the love towards her brother. Antigone had no doubt: what counts more is the latter. Both Creon and Antigone did put their (different) Personal Philosophies under scrutiny. They both used them dogmatically in the process of the choice: as an easy and unquestionable decision-maker at their disposal. However, at a certain moment, something happened. Due to the plea of Tiresias, the blind prophet, and the complaints and sorrows of the Chorus, Creon changed his mind. Unfortunately it was too late! Antigone had hanged herself and Haemon had stubbed himself. And even Eurydice—Creon's wife and Haemon's mother—committed suicide. It was too late to change Creon or Antingone's preferences and therefore to avoid the bad consequences of bad decisions taken stubbornly on the bases of certain beliefs, belonging to their respective Personal Philosophies, considered unquestionable. When Tiresias and the Chorus spurred Creon to be aware of the situation, they were actually offering a rough and not elaborated *Ethical Counselling* and what we have called a *reflection slot*. Thanks to this, the King changed his mind. But this intervention occurred, as said, too late. He should have received a proper counselling before. He should have had a critical reflection slot before. Probably he would not have changed his mind, or probably he would. But in any case he would have chosen with more awareness of the situation and of his beliefs and values.

While in Sophocles' *Antigone*, the two different possibilities are "embodied" into two different individuals, Creon and Antigone, in Aeschylus' *Agamemnon*, the two are "embodied" into the same individual. Agamemnon had to choose between two options: (i) the life of his beloved daughter Iphigenia but the infringement of Gods' will to sacrifice her for the good of the incoming war against Troy; (ii) the obedience to Gods' will but the death of Iphigenia. Agamemnon knew that any decision he would have taken carried bad consequences. However, he analysed the situation and then decided.

Sophocles' *Antigone* teaches us that there should be a reflection slot and that there should be the way of critically thinking about the decisional situation in which we are by "fortune" and about the Personal Philosophy we possess. Aeschylus' *Agamemnon* teaches us that many times our decisions might turn into bad consequences not only for us but even for our relatives and beloveds. However, this awareness cannot put us into a decisional paralysis. At a certain moment, we have to decide, and if we are not able to decide by ourselves, others could consciously or unconsciously impose their viewpoints, and therefore their choices, on us.

The real Giovanna, Claudio, Anna, Julie, and Veronica had to decide, and they might have done this simply by using their Personal Philosophies, without posing them under critical scrutiny and considering them as an untouchable dogma. They could have turned out into a decisional paralysis and asked for an external decision, without wondering whether this decision was really the decision they wanted to reach. But they could have demanded for an Ethical Counselling. This would not have surely produced a *prêt-à-porter* solution for their problems, but it would have developed a process through which they could have reached greater awareness of their situations, so as to individuate their values, their beliefs, and even their dogma.

Summing up, proposing an Ethical Counselling means proposing *patient first*, i.e. patients' quality of life and the imbricated quality of their decisional processes, especially now in the age of personalised medicine.

References

Nussbaum M (2001) The fragility of goodness: luck and ethics in Greek tragedy and philosophy, 2nd ed. Cambridge University Press, New York
Williams B (1993) Shame and necessity. University of California Press, Berkeley

Printed in the United States
By Bookmasters